Managing E-Business Projects

By

Wes Balakian PMP

Keith Young CISA, CePM

Rajesh Veerapaneni

ISBN: 0-7596-8472-3 (Ebook)
ISBN: 0-7596-8473-1 (Softcover)
ISBN: 0-7596-8474-X (Hardcover)

Library of Congress Control Number: 2001098403

This book is printed on acid free paper.

Printed in the United States of America
Bloomington, IN

1stBooks - rev. 4/25/02

Table of Contents

Chapter 5

Chapter 6

Chapter 7

Chapter 8

Designing the eBusiness Solution Architecture 86

Chapter 9

Building and implementing an eBusiness System 98

Chapter 10

Securing Your eBusiness Application 117

Acknowledgements

Without question, this book could not have been written without the support, inspiration and help of our wives (Lorry Balakian, Gina L. Young and Jyothi Veerapaneni). Words cannot express how grateful we are for your support and dedication.

We would also like to thank Lori FredeKing for helping us organize and document our thoughts. Lori, your support helped us stay focused. Finally, we would like to thank Mike Price of the Project Management Institute for his support and encouragement. Mike, your willingness to give us a chance to be pioneers in the area of eBusiness project management is greatly appreciated.

Preface

Guide to the reader

For those of you who are ready to get the practitioners perspective of eBusiness project management from three subject matter experts, you have come to the right place.

This book is intended to be a reference for those of you who have some experience in project management and are not seeking a methodology and process manual. For those of you with some eBusiness experience and wanting to gain insight, direction and sound practices for managing eBusiness projects and would like a good reference and validation of how to deliver your "e" project, then read on. For the many who are just curious to know what everyone is talking about regarding eBusiness project management and what there is to know about delivering an eBusiness project, then you too should find many answers to your questions within this book.

For the few who read this book out of curiosity or possibly seeking to change their careers, we trust that you find insightful and practical knowledge about project management and eBusiness implementation from a project managers perspective.

That being said, to the rest of the readers of this book, we have made some assumptions, throughout this book we hope that you will consider those assumptions as you read our work. We assume that you have a good foundation in general business practice, at least some knowledge about the Internet and as a minimum a solid introduction in project management philosophy.

Just as others may have done in this exciting profession, we did not wake up one morning and announce to the world that we had "automagicly" (technical term) became eBusiness experts overnight. All three of us developed our skills, experience and knowledge from other beginnings. All of us in some form or another have spent time in IT (Information Technology), IS (Information Systems) and technology management in one form or another. All of us have specialized in various industry specific areas like, risk management, security and procurement.

What we feel made the difference is in the very beginning, and we mean the very beginning, all three of us realized the potential of what the Internet could and will be. We all realized what it could mean for business, some of us saw this potential for business to customer relationship before realizing the true value to traditional business to business relationships.

Each of us was able to visualize what business would be like if it were truly conducted through the Internet and beyond. This vision that each of us carries inside and comes out every time we start to talk about the subject, is what has driven each of us to where we are today.

That is the short story behind how we have developed our skills, trained our peers and now want to share our experiences with you. For the rest of the story and some helpful concepts, thoughts, tips and techniques, read on.

Chapter 1

eBusiness: Where Have We Been and Where Are We Going?

Chapter Preview

The Internet has undergone several changes. In the late 1990s, the Internet was primarily used for sending and receiving electronic messages and creating brochure ware-type applications. Today, companies are using the Internet to extend their internal processes, protect their information assets, and improve the overall management of their enterprises. Companies are also using the Internet to gain competitive advantages with new business models that target e-marketplaces, such as B2B marketplaces and online trading hubs.

Your role as an eBusiness project manager, or EPM, has also evolved with the changes in the Internet environment. You will now be required to manage the implementation of your organization's eBusiness solution through traditional duties, such as delegating authority, controlling project costs, and recruiting and leading project personnel, but you will also need to raise your level of thinking and performance to include a deeper understanding of your organization's business strategy, its customers and business partners, and how the Internet should be used by your organization to create value and a competitive advantage.

This chapter will provide a short definition of the term "eBusiness"; an overview of some of the Internet business models that are being used by companies to conduct business online and discuss some of the elements that require analysis when selecting an Internet business model for your company or project. It will also examine the recent eBusiness market trends and how they have impacted the business community and explore the future of the eBusiness market space.

eBusiness versus eCommerce

The terms eBusiness and eCommerce are often used to describe online business activity. Though both terms are very similar, there are some differences worth noting. The term eCommerce refers to the integration of Internet technology into only customer facing processes of an organization, while eBusiness refers to the integration of Internet

1

technology into the customer facing and back end processes of an organization.

An example of eCommerce is when a company creates an online form for customers to place orders over the web. Conversely, an example of eBusiness is when a company creates an online form for customers to place orders and develops online reports that are accessible by individuals in the billing departments to reconcile customer payments and billing invoices.

There are several reasons why the EPM should understand the difference between eCommerce and eBusiness, but the most important are that this knowledge will help in better understanding the scope/requirements of the project and identifying stakeholders (internal and external) that will be impacted.

eBusiness Models

What is an eBusiness Model?

The term, "ebusiness model," is often misused. It is often used to describe how a company plans to compete in a specific marketspace, but it is also used to describe how a company will re-engineer its internal operations to support its eBusiness strategy, such as accepting customer orders, shipping products, and collecting payment. If you are new to the field of eBusiness project management, understanding the meaning of the term, "eBusiness model," can be confusing.

You should understand the difference between an eBusiness strategy and an Internet ebusiness model. An eBusiness strategy represents the goals that have been set by a company regarding the use of the Internet, such as improving customer service by accepting customer inquiries via the Internet. An eBusiness model, on the other hand, represents the processes and activities that are to be performed to attain the goals of an eBusiness strategy, such as the creation of a customized Web site to permit a service representative to provide customer support over the web.

Types of eBusiness Models

There are a several types of Internet business models used by organizations. Each has its own unique characteristics. Some of the Ebusiness models in use today include:

- Business to Business (B2B)
- Business to Consumer (B2C)

- Consumer to Consumer (C2C)
- Business to Employee (B2E)
- Consumer to Business (C2B)

Business to Business (B2B)

A B2B model involves the use of Internet technology by two or more companies to conduct commerce. This business model has received significant attention from investors and the media alike. According to leading research firms, companies that competed using a B2B model generated over $50 billion worth of commerce in 1999. An example of a B2B model would be an automotive parts supplier that implements a web site to allow Buyers at original equipment manufacturers (OEMs) to order its products.

In addition to the traditional B2B model, where only two companies engage in commerce (the exchange or buying and selling of goods, services or commodities), there are four other B2B models worth mentioning:

The trading hub model, which involves the deployment of Internet technology to provide business-to-business commerce opportunities for companies that operate within multiple vertical markets.

The auction market model involves multiple buyers and sellers competitively bidding on a contract.

The post-and-browse model, which involves buyers and sellers posting their procurement or selling interests on a bulletin board.

The aggregation model, which involves a company streamlining purchasing process by aggregating the product catalogs of multiple suppliers in a centralized environment using a standard product catalog format.

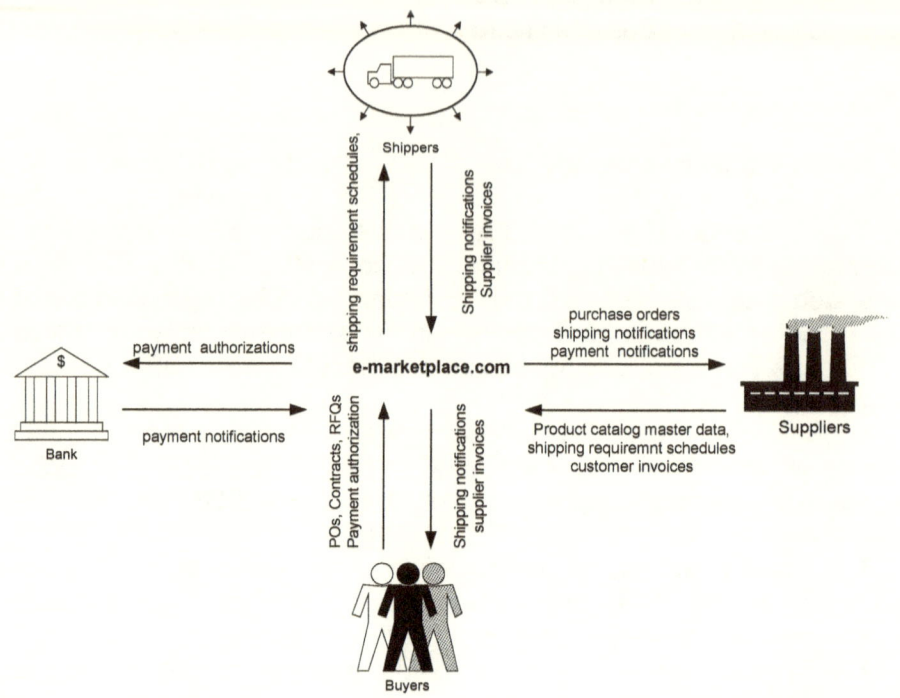

Figure 1.1 B2B Model

Business to Consumer (B2C)

The Business-to-Consumer (B2C) model is the most mature of all the eBusiness models. Its architecture is designed to extend the front-end operations of a business out to the individual consumer. One key difference between the B2C and the B2B model is that the B2C model does not allow for price and market activity transparency, and it uses more of a "top-down" delivery approach, meaning that the company controls more of the relationship.

The use of a B2C model is often found in Internet commerce retail applications when a company offers its product to individual consumers via the Internet with credit card or e-check payment options.

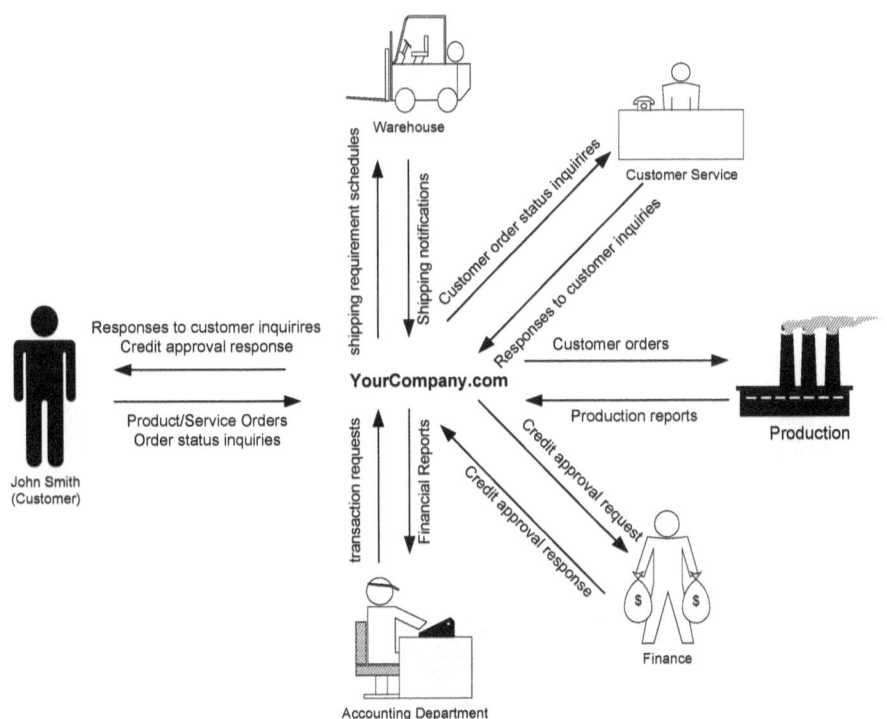

Figure 1.2 B2C Model

Consumer to Consumer (C2C)

The Consumer-to-Consumer (C2C) model is designed to provide individuals the opportunity to conduct commerce on a one-to-one basis. The value of this business model is that it extends the control of conducting commerce down to the individual level, where the individual buyers and sellers can control various elements of the commerce transaction, such as price and product specifications. Examples of companies that use C2C model include eBay.com, a leading online auction service provider, Flycode and Distributed Science Corporation (California-based companies that allows users to swap digital content using Peer to Peer (P2P) networking technology).

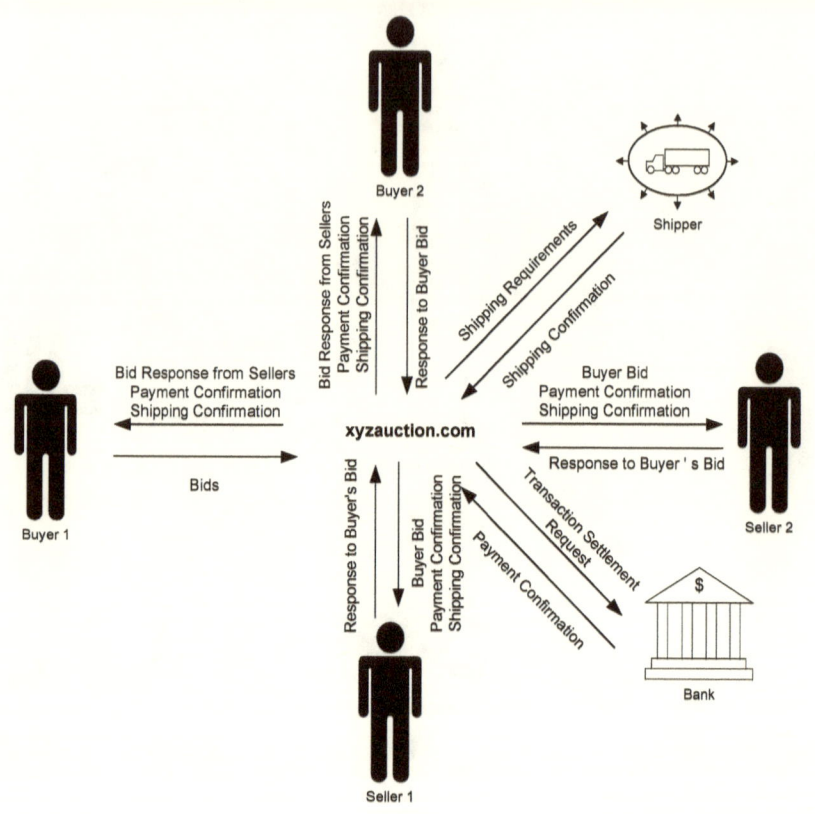

Figure 1.3 C2C Model

Business to Employee (B2E)

The Business-to-Employee (B2E) model involves the use of the Internet to streamline systems and processes that are used by the employees of a company. It is very similar to the B2C model. The key difference is in the characteristics of the end consumer. In a traditional B2C model, the end consumer does not have a business relationship with the seller, such as being a business partner or an employee, while in a B2E business model, the end consumer is either employed by the seller or has a pre-existing relationship. Some reasons to implement a B2E model may include to reduce the cost of communicating with employees, improve the productivity and efficiency of the internal operations of the company, and increase the revenue stream generated from "internal customers."

An example of a company that uses the B2E business model is IBM Corporation®, which offers its computer products, sometimes at a special price, to its employees.

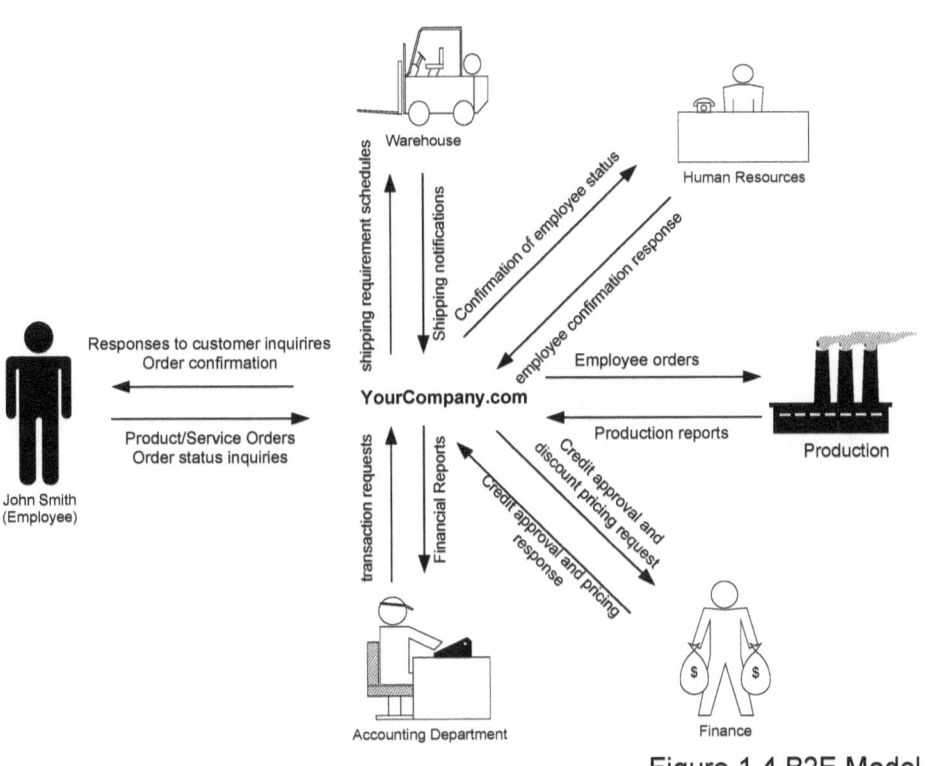

Figure 1.4 B2E Model

Consumer to Business

The Consumer-to-Business (C2B) eBusiness model involves the use of Internet technology to enhance the relationship between the buyer and seller by empowering the buyer. In this business model, the buyer drives the relationship by quoting a price that they are willing to pay for a certain product or service. The seller, in turn, accepts or rejects the buyer's offer.

In the auction business, this type of activity is considered a "reverse auction." An example of a company that uses the C2B model is

Priceline.com, which lets users submit price offers to the company's suppliers. Although this business model creates value for the buyer, it has not received as much acceptance by the eBusiness community as other business models have, because of the amount of effort required on the supplier's side to control cost and maintain profitability.

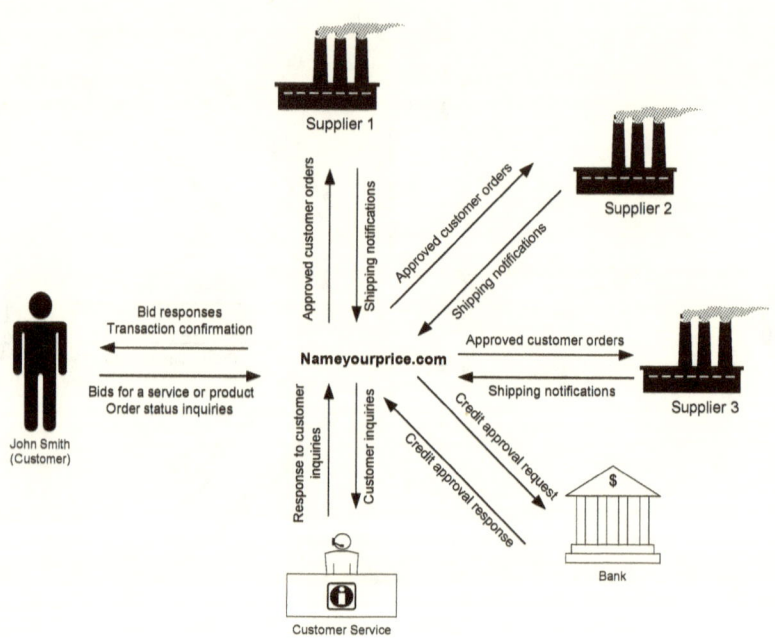

Figure 1.5 C2B Model

Selecting The Right eBusiness Model

Even though any of the previously discussed eBusiness models can be used to enable a company's eBusiness strategy, one of the keys to successful use of the Internet is having a good understanding of which eBusiness model to use. This concept is important, because the use of one business model over another will affect the complexity of your project and impact on other project management-related issues, such as resource requirements, budget requirements, and scope management.

For example, your client organization may decide to enable its business strategy by employing a B2B eBusiness model, rather than a B2C eBusiness model. If the B2C model had been selected, the scope of your project would have only included an understanding of the consumer experience on the front-end and the integration requirements for your organization's back-end systems. Because the B2B model was chosen, the scope of your project has just doubled. Not only are you responsible for the integration of the eBusiness solution within your organization, but you are also responsible for ensuring that the solution integrates seamlessly with the internal components of your organization's business partners and coordinating with a larger project team. You now must find a way to share information with the project staff that is employed by your organization's business partners. This may sound simple initially, but it's not uncommon for complications to arise, such as members of the project staff to be located three time zones away.

In addition to an increase in scope and resource management requirements, the amount of effort required to procure vendor products and services has just become more complex. When using a B2B eBusiness model, the complexity of the technologies are greater than the complexity of technologies that are used to enable a B2C eBusiness model. For example, a company, looking to implement a B2C model by deploying a simple storefront with back-end connectivity to a credit card processor, can do so by purchasing any one of several off-the-shelf software applications, such as Microsoft Frontpage®. On the other hand, a company looking to deploy a B2B exchange for the health care industry will need to purchase applications that are customized for the healthcare marketplace, thus accounting for that industry's business rules and regulatory requirements.

Selecting the "right" eBusiness model can have a significant impact on your project. Even though the project manager is often not consulted when the business model is selected, having a basic understanding of the elements considered and the process involved in selecting a business model can be an asset.

The selection of a business model involves a detailed process where the business strategy, performance time lines, technology and project resources are analyzed. The focus of the analysis process is to select an Internet business model that will enable a company to reach its strategic goal in the shortest amount of time with the least amount of investment (money, people, land, equipment), while using the most effective technology solution.

For example, a retail sales company would like to use the Internet to increase its market share of consumers that live outside the United States by 20 percent in just 90 days, but the company only has $5,000 available for the project. Even though the company's funds are limited and its time line is short, it could theoretically achieve its goal through the creation of a "Web presence" with a B2C model by setting up a storefront with an Internet mall focused on the international marketplace. However, it would probably not be wise to create a Web presence using a B2B model (e.g. joining an Asian-based marketplace), because the amount of money available to cover the cost of the project and the project implementation time line.

In addition to analyzing the business strategy, project implementation time lines, technology complexity, and project resources, the following would also need to be considered:

- Market size
- Profitability
- Application customization requirements
- Technology deployment and maintenance requirements
- End-user support requirements

Market Size

Determining the market size is important, because it will provide you with information regarding the potential liquidity, in dollars and number of customers, of the marketplace in which your company will compete. Performing this task takes time, effort, and access to multiple information sources. One of the ways to determine the market size is to gather information from leading market research firms. For example, according to the Yankee Group, a leading market research firm, companies that compete in the B2B marketspace will conduct over $3 trillion in commerce via the Internet by 2003, while companies that compete in the B2C marketspace will only conduct $108 billion in commerce via the Internet over the same time period.

Another way to determine the market size is to analyze how your competitors are using the Internet. This is an effective approach, because it will not only provide you and your company with a snapshot of what the competition is doing, but it will also give you a chance to identify weaknesses in your competition's approach. For example, your company's competitors may have the same eBusiness strategy, such as increasing its market share in the business-to-business sales market, but it may be using a different Internet eBusiness model to enable its eBusiness strategy, such as by implementing a horizontal B2B exchange versus deploying a market-specific trading hub.

Profitability

Profitability is a key factor when selecting an eBusiness model. Without a consistent stream of cashflow and profitability, your company will not survive. Some of the ways to determine whether or not your eBusiness model will be profitable is to examine the potential profit margin (gross sales price of the product or service being sold less the cost of goods sold). Based on the eBbusiness model chosen, the profit margin can vary significantly. For example, a company using a C2C Internet business model may have a lower profit margin than a company that employs a B2B model due to the model chosen.

Application Customization Requirements

Customizing an application can involve significant time, effort, and expense. For a company that is seeking to implement a B2B eBusiness model, the application customization requirements can be significant. For example, if a company is deploying a B2B exchange that will support the chemical industry, customization may be required for procurement and shipping activities that involve hazardous materials. It may also require customization of back-end systems in order facilitate sharing of information between the market participants.

Technology Deployment and Maintenance Requirements

Performing an analysis of the technology deployment and maintenance requirements is also important to consider when selecting an eBusiness model, because it can impact your company's profitability and its ability to meet the dynamic needs of its customers. For example, in a B2C model, the deployment and maintenance requirements for the technology infrastructure may be lower than those for a B2B model. Ongoing maintenance is often the most overlooked cost of implementing an eBusiness solution.

End-User Requirements

End-user support requirements involve the amount of resources and effort required to support the users of the application (e.g. customers). Understanding this requirement will give you an insight into what additional costs could be incurred to deploy the business model. For example, a company deploying a C2B model may hire additional

11

customer support staff and implement other customer-facing support processes, even though the end-user support requirements may only really need to include retraining existing customer support staff.

eBusiness Market Trends

The Internet market has experienced extreme trends. Some the resulting trends in the Internet economy have included:

- Financial problems with dot-com companies.
- Development of eBusiness project management discipline.
- Growth in legislation used to protect consumer privacy and security.
- Development of advanced Internet technologies.

Financial Problems with Dot-Com Companies

From an economic perspective, the Internet economy has been on a roller coaster ride. In 2001, several eBusinesses closed down, while others were on "financial life support." During the last quarter of 2000, the NASDAQ, a leading stock exchange for trading technology stocks, saw significant declines in the prices of technology stocks, and subsequently, the decline in market value of many of its top Internet companies. The downturn in the stocks of Internet companies was not a sign of a weak industry, but more of a maturing process that every new marketplace goes through. The only difference is that the Internet economy is maturing at a much faster rate than most markets of the past.

The most notable trend in the eBusiness marketspace has been the decline in Internet-based companies. According to an article in the January, 2001 edition of *Informationweek*, nearly 210 dot-com companies closed down during 2000 and over 15,000 jobs were lost. Of the estimated 210 companies that closed down, 75 percent were B2C companies, while 21 percent were B2B companies and the remaining 4 percent consisted of online services and infrastructure businesses.

The decline in the eBusiness marketspace was partly due to the natural maturity process that takes place with any new market, but it can also be attributed to poor management. The outlook for the Internet economy on the surface seemed pretty bleak, but for those companies that have a good business strategy, strong management team, and access to capital, the Internet still is, a great place to be.

Development of the eBusiness Project Management Discipline.

The development of an eBusiness project management discipline has been one of the bright spots throughout these changes. eBusiness project management is important, because it provides project managers with the skills and knowledge to effectively manage the implementation of an eBusiness technology, while at the same time advancing the profession forward. The increased interest of project managers in the eBusiness project management discipline can be attributed to the number of training programs and "certifications" that are being offered by universities and professional associations and the number of Web-based project management tools that are now available.

The eBusiness project management discipline is in its infancy. As companies increase their use of Internet technologies, it will add new business models and strategies to the skill set and knowledge requirements of the eBusiness project manager. In addition to an increase in the eBusiness project manager's knowledge base, the position of project manager will evolve to more of an executive position, where the project manager will be able to become more active in developing the business strategy for their organizations. (For more information on the role and responsibilities of the eBusiness project manager, read Chapter 2.)

Growth in Legislation to Protect Consumer Privacy and Security

There has been a significant increase in the development of legislation by regulatory agencies to protect Internet users. The growth in legislation from organizations, like the Federal Trade Commission and the Department of Human Health Services, has been squarely directed at protecting consumer privacy and security. The growth in this legislation has been primarily due to Internet users' concerns over some eBusinesses' abusive practices employed to gather information on customer preferences, like the unauthorized use of cookies to track what Web sites a user visits.

Some of the legislation that focused on privacy-based issues include the Children Online Privacy Act, Health Insurance Portability and Accountability Act, (HIPPA) and the Financial Information Privacy Act. Legislation was also passed to protect intellectual property rights and establish guidelines for securing Internet applications. (To help ensure that your eBusiness solution is "safe," read Chapters 7 and 8, which cover privacy, security, and intellectual capital protection issues in more detail.)

Development of Advanced Internet Technologies

Advancing technology has improved the functionality of eBusiness applications. In the summer of 2001, there was an increase in interest and money invested in P2P networking, Public Key Infrastructure (PKI) security, and wireless communication technologies, especially Wireless Access Protocol (WAP). P2P technology allowed users to share digital content files via a special communication protocol, while PKI security is being used to enhance the security of Internet applications by providing better techniques for authenticating users and protecting data. Wireless communication technology is being used to transmit information between users of Web-enabled devices. Wireless Access Protocol, or WAP, is a communication protocol used to enable devices not physically connected to the Internet, such as cell phones and I-appliances, to send and receive information to other Web-enabled devices.

Several other technologies have also been developed to extend the functionality of eBusiness applications, such as Extensible Markup Language (XML) and digital rights management technology. Extensible Markup Language improves the ability for computers to share information. Digital rights management technology protects intellectual capital and improves the authorization management techniques used to give users access to a company's information assets.

Future Direction of The eBusiness Market

The eBusiness market is constantly changing. Its continued maturity is primarily related to companies finding new ways of using Internet technology and the growth in the capabilities and functionality of Internet applications. In the near term, the eBusiness market will continue to go through the "cleansing" phase, where companies with poor business models, management teams, and market positions will cease to exist and close operations. Following this period, the companies that remain will find ways to further refine their operations through the use of the Internet. For example, many eBusiness solutions today lack full back integration with legacy-based systems, but in the very near future, this will no longer be a problem. The solution will come from both technology vendors, such as BEA, i2, and Weblogics, and business professionals that understand these issues and what is needed to solve them.

In addition to the above, advanced technology architectures will be developed to further reduce the time and cost of deploying eBusiness solutions. (For a detailed analysis of the various technology architectures, read Chapters 11 and 12.)

Chapter Summary

The eBusiness marketplace has changed significantly. In this chapter, we defined the terms eBusiness and eCommerce; examined the different business models being used by companies to conduct commerce over the Internet. We also analyzed factors to consider when selecting the "right" Internet business model: market size, profitability, application customization requirements, technology deployment and maintenance requirements, and end-user support requirements. We examined some of the top trends in the eBusiness marketplace: the financial problems in the dot-com industry, the growth in the development of eBusiness project management discipline, the growth in the development of governmental legislation to protect consumer privacy and security, and the development of advanced Internet technologies.

By covering these topics, you should now understand some of the unique elements involved in being a project manager of an eBusiness project. With this knowledge, you are now able to move forward to Chapter 2, where you will learn about the evolution of and differences in the roles of a traditional project manager and an eBusiness project manager, the definition of the role of an eBusiness project manager today, and the tasks that an eBusiness project manager must focus upon.

Chapter 2

eBusiness Project Management Overview

Chapter Preview

This chapter will first explore the evolution of and differences in the roles of a traditional project manager and an eBusiness project manager. It will then clearly define the eBusiness project manager's role in the implementation of an eBusiness application as it has evolved to today. With a foundation built in role delineation, this chapter will then identify key tasks that an eBusiness project manager must focus on to be successful.

As we move through the processes and practices in eBusiness project management, you will discover how much deeper and wider the enterprise of the eBusiness project manager must go to be successful in today's market space. The more an EPM understands about the different management skills and business acumen needed to operate successfully in the Internet environment, the more he or she will realize that the role of an eBusiness project manager is being elevated to an executive-level position.

Evolution of Traditional and eBusiness Project Managers Roles

Early in the evolutionary chain of the Internet, buyers and sellers could transact business through a closed and very expensive, non-scalable system using Electronic Data Interchange (EDI). EDI was a good model (or messaging system) as long you had enough money and resources to maintain the system. Buyers and suppliers were the only ones engaged in this closed loop network. More recently, brochure-ware and B2C (Business to Consumer) networks began to emerge using a new medium, the Internet. Basic e-commerce began to gain in popularity. Using the Internet enabled a one to many relationship which is when e-commerce transactions began to accelerate. One to one selling gained momentum and trust beginning in 1996 and carried on through the end of 1998 and the beginning of 1999. As B2C matured a new model emerged, called B2B or Business to Business. At this stage, using the B2B model, market efficiency began to accelerate rapidly. As buyers and suppliers were able to conduct business at a blistering pace using web technologies and security that allowed a many to many relationship that facilitated business among many suppliers and buyers. This model is what many project managers are familiar with and work with today.

All of this transformation during such a relatively short time, give or take ten years, has created a high demand for competent business savvy project managers with knowledge in business operations, software development, web technologies, security and transaction processing with a focus on project management methodology needed to make these types of projects successful. As technology continues to change and mature, the requirement to have highly trained, technically sound, project managers involved in mission critical web based and business to business projects grows as well. More and more companies are relying on their existing project teams to be able to execute eBusiness applications in record time with little room for error. Many of these teams have succeeded, some as you know have not.

As a professional discipline in more recent times, project management has been practiced since the early 1960s. Today, project management, as defined by the Project Management Institute®, is "the application of knowledge, skills, tools, and techniques to project activities to meet the project requirements. Project management is accomplished through the use of processes."

Although, in terms of scope and history, the practice of eBusiness project management is a very new discipline, it is all of the above and more. When a business determines that it is going to pursue an eBusiness solution or decides to convert a segment of business into eBusiness, many decisions are precipitated, such as which technology is best, how to promote the program in the context of a phased approach or an all-at-once delivery scenario, how this aligns with the existing business model, how the post implementation will be handled, etc. Therefore, the use of existing standards and processes of project management is only a small piece of the overall scheme of what it takes to manage an eBusiness project.

Traditional Project Management vs. eBusiness Project Management

Understanding that there *are* vast differences between traditional project management and eBusiness project management is paramount to becoming a successful e-project manager.

In Figure 2.1, some of the differences between traditional IT development projects and eBusiness projects are highlighted. (The reason for the "unknown" in the last row of the third column is because

17

the new project management components that will be needed in the future is, indeed, unknown. Will they be wireless commerce, remote project management, or m-commerce skills?)

Figure 2.1.

Differences in Traditional and eBusiness Project Management

Project Area	Traditional Projects	EBusiness Projects
Speed	Systematic	Rapid
Duration	Long	Short
Implementation	Methodical	Quick
Skill set	Taskmaster	Change agent
Scale	Large scale, long duration Independent	Smaller scale, increasingly Interdependent
Visibility	Behind the scenes	High profile
Cross organizational role	Small, shallow	Wide, deep
Communication	When needed	Constant, distributed
Customer expectation	Market driven	Customer collaborated
New Components	EBusiness	Unknown

Roles and Responsibilities of an EPM

As you can see, the skill set for eBusiness project managers is different from the skill set of a traditional project manager's. A traditional PM is not always, nor often, the decision-maker because most of the business, technical, functional, and nonfunctional decisions are typically made by team members responsible for their respective part of the project. However, an EPM, although he or she may or may not perform many of these functions, must be aware of all tasks that need to be completed to finish the project and whom is responsible for completing these tasks. To do so, an EPM must communicate and distribute real-time information, coordinate, anticipate, and lead the project team, while mitigating any risk as may be determined by the team. As the leader or conductor of a project, an EPM's role is to guide the project team, while making business decisions that are clearly achievable, technologically sound, and customer centric in order to achieve profitability, customer acceptance, and the strategic objectives of the company. The EPM must lead the project team to the ultimate completion of the project in every aspect.

Not only does the project information need to be communicated within the project team, but also with other organizations within the company. As an EPM goes deeper and wider into an organization, he or she must use their management skills and knowledge of the business to meet the *customer's* needs. If your customer or end user is within the company you are working for, the project manager plays the role of managing the change throughout the many organizations or separate business units that will be engaged by the decision of going to an Internet commerce solution. This is will be validated by the many different organizations that will become involved with an e-project, such as legal, marketing, Information Technology, graphics design, shipping, procurement, new product development, accounting, and finance. Thus, the eBusiness project manager's role has become one that is much more involved in the business decision process and more of a cross-organizational function. This new role is sometimes described as that of a change agent.

For example, in the telecommunications industry, just about everything is a project. A certain level of support, integration, and cross-organizational involvement is required to get projects completed. Additionally, mainframe projects have evolved over the years from a behind-the-scenes process to a more customer-focused, high-profile endeavor for a project manager. As the complexity, costs, and duration of projects grew, so did the need for formal organizational planning and control, especially tracking and quality measures. Project risk, cost and impact became of ultimate importance, so it became required of an eBusiness project manager to be able to maneuver through the complexities of a corporate organization with cunning ease, be versatile enough to work with any business unit to complete the tasks at hand, and have communication skills that were effective from the project war room to the boardroom.

Each of these roles has a direct impact on how the project manager will perform. For example, if a project manager's skills are very technical in nature, he or she may have a tendency to support and over manage the areas in which he or she is most comfortable, such as software functionality and features, and not focus on other equally important areas of the project, such as communications, planning, risk management, implementation, and content management. It is natural for managers (and for that matter, most people) to drift into their comfort zone, where they feel they can provide the most impact by virtue of their own knowledge and experience.

The best project manager is one that will seek out the areas in which they are most unfamiliar, such as leadership style, technical competency, business knowledge, and communication skills, and begin to understand those areas to the best of their ability. Effective leaders recognize their comfort zone and frequently step out of it to enhance their competency.

Skills of Successful EPM

Managing EBusiness is about managing change. As we mentioned earlier, one way for you as an EPM to prepare for an EBusiness project is to be prepared to manage change. Not only should you be willing to accept change but you must be able to manage change in order to succeed.

Business conditions change rapidly when business is conducted over a global electronic network that connects all business participants from the manufacture to consumer. When your project goes live you seldom (if ever) get a second chance to leave a good impression. One of the most exciting challenges in the eBusiness world is the fast changing face of competition. Keeping up with your competition takes on a different perspective as an EPM. Your skills in improvising and creativity will be tested as the deadlines for go-live launch dates shrink, competition pops up from nowhere and internal pressures to perform within the triple constraints of cost, quality and time mount.

Why then, have many of us decided to pursue this exciting and rewarding new profession called EBusiness project management? The authors of this book have experienced first hand what it is like to be an EPM in various environments and settings. Each of us have been fortunate in that we have been through the storm and can see the clearing ahead. Because others like ourselves are sharing their experiences and lessons learned on this subject and because organizations are recognizing that an EPM has a different skill set,

20

approach and technique when it comes to delivering an Internet enabled solution, the discipline of becoming an EPM is advancing and maturing.

What are some of the skills of a successful EPM?

- Communication - Both upward and downward as well as across the organization.
- Listening - From a project team perspective and from a stakeholder perspective.
- Understanding the business - Must have a good understanding of the business and how it operates.
- Understanding technology - Must be able to communicate with the technical team.
- Attention to detail and a strong sense of organization are obvious skill requirements.
- They require strong logic and analytic skills.
- The ability to ask appropriate questions and communicate effectively to a broad range of people.
- Must be able to conceptualize the abstract and manufacture the concrete to explain it.

Ways to Fail as an EPM

- Take a hands-off approach to project administration
- Let issues drift and remain unresolved allowing them to fester
- Be unwilling to listen to suggestions for change
- Be over focused on specific project management tools
- Devote too much attention to relations with management and not enough to the project team
- Attempt to micromanage the project and not delegate
- Be willing to rapidly adopt new tools without assessing the consequences
- Be tool-focused as opposed to method-oriented with the tools supporting the methods
- Fail to regularly communicate in person with all key members of the project team

Issues and Challenges of eBusiness Project Management

One of the biggest challenges facing business in the future is how to meet the needs of changing business drivers. For the near term, many of those changes will be technology driven. As technology changes, so does the need for changing skills. Anyone who learns to adapt to the changing landscape of business will be better prepared to face the challenges these changes bring. As an EPM you will be one of the few who will be prepared for these changes. What you need to know as an EPM will be outlined through out this book. Each e-project you work on will be different in some way. Use every failure, and you will have failures, as the springboard for success.

Keeping projects on track

One of the hardest issues facing companies today is getting a product to market in a timely manner. Idea-to-market turnaround time has decreased over the past few years. A product or service has to be delivered quickly if it is to be successful, especially when the product or service is Internet hardware or software related. It has been said (by Internet research companies and published in CIO magazine) that if you cannot introduce your Internet based product or service within two weeks of your competition, you have lost over fifty percent of the market share. Conversely, if you are first to market you have gained that market share and a competitive edge over your competition. What role does the project manager play in the success of this effort? The project manager is the one person who is able to look across the company's organizational components and rally the various executive sponsors to focus on the task at hand. Not only is there a need for this coordination, but also the need to manage to a schedule and keep the entire team focused and motivated to meet or exceed deadlines. Without focused guidance, communication and management of the project, you have many talented people all working within their various areas of expertise but not moving in the same direction. Their focus is to keep the right people moving in the right direction at the right time, so that the entire project objective is met. By always looking forward and anticipating the outcome, the project manager's goal is to keep the team on course. In good and bad economic times there must be a way to balance the work load for companies. There must be experts who can quickly come into a company and asses their needs regarding an eBusiness application. Enter the EPM.

Outsourcing

There are alternatives to the do-it-yourself approach. A wide range of service providers of different types and capabilities offer assistance with the arduous task of moving from traditional business to eBusiness. These companies can step in and take charge of some or all of your IT and network infrastructure needs or some portion of a specific implementation. Some of the companies available today can offer service as wide as a complete end-to-end solutions including hosting and post-production support. Other companies offer boutique services such as specialized programming, content management and development, or simply project management support.

By focusing on its business—specifically managing the infrastructure which underlies its clients' applications—an outsourcer can free a company's in-house staff to stay focused on core business issues and bottom-line goals. Done right, outsourcing delivers improved application uptime that translates to revenue growth and greater customer loyalty and retention—usually for a controlled, predictable fee.

Buyer beware: Not all outsourcing is created equal. Many first-generation hosting providers offer a piecemeal infrastructure comprised of technology components that cannot keep pace with the demands of eBusiness. Typically, their customers are left with the burdens of not merely application development, but also integration, implementation and equipment provisioning. Be selective and do your research *before* deciding on an outsource partner. They must have a stake in the project. If your company does not have a contract management organization, then you must make every effort to ensure that your selected supplier is going to deliver what you expect when you expect it.

In today's marketplace, across all industry segments businesses are realizing that transformation to eBusiness is required to remain competitive. Analysts predict that companies not making the necessary changes will be overrun by competition and ultimately fail. As enterprises around the world undergo this transformation, they are increasingly leveraging outsourcing options to help:

- Broaden their markets by extending their reach globally.
- Enter new business areas through collaborations or expanded services made possible with web-based interactions.

- Increase employee productivity by providing easier access to corporate information and services.
- Reduce costs through improved operations that integrate web access and traditional information technology (IT) systems.

Chapter Summary

This chapter explored the evolution of and differences in the roles of a traditional project manager and an eBusiness project manager and defined the eBusiness project manager's role in the implementation of an eBusiness application as it has evolved to today. We discovered that the project manager is also a change agent, not just a taskmaster. With a foundation built in role delineation, you are now ready to explore the key tasks that an eBusiness project manager must focus upon, beginning with Chapter 3—Selecting a Software Development Life Cycle.

- Planning the project
- Managing stakeholders and the project team
- Defining business requirements
- Designing an eBusiness system

Where do we go from here?

The following pages outline the remaining chapters in summary and as reference to quickly locate specific subject matter. As a project manager yourself we assume that you have an understanding of project management, the Internet and generally accepted business practices. As you read the remainder of this book, our approach to delivering information that is practical, applicable and relatively short on theory, can also outline the approach of an EPM. Communicate clearly, document the details and deliver on time what you agreed to, are all traits of an good EPM.

Selecting a Software Development Life Cycle

Chapter 3 will explore seven of the most popular software development life cycle models and the specific advantages and disadvantages of each model and how each model may be used in various applications to meet a project's objectives and goals. As you review each model, be sure to focus on the model itself, rather than on the method and unique application of a particular project, and how to best use these models to meet a project's objectives.

Defining the e-strategy

Chapter 4 will explain how the development of an e-strategy begins by identifying business processes that can be e-enabled. During this process, an e-project manager must assess how the e-strategy will impact the organizational culture, productivity, and efficiency, then determine the potential value contribution for an e-enabled application as it may apply before, during, and after deployment. Additionally, this chapter will also explore the factors to be considered when selecting an implementation model, provide a sample implementation table for identifying these factors, and discuss the option for an EPM to outsource or develop the applications in-house.

Planning the project

Chapter 5 will discuss many of the different aspects of planning an eBusiness project, such as the steps involved in managing risk; constructing a project charter; selecting and working with vendors; funding an eBusiness project, including identifying, estimating, and controlling costs; and developing a Work Breakdown Structure.

Managing Team Personnel and Project Stakeholders

To assist the EPM with managing his or her most valuable resource, Chapter 6 will discuss the areas of assembling an eBusiness project team, setting up a team architecture, identifying and documenting key skill requirements, identifying and communicating with project stakeholders, and keeping the team personnel, thus the project, on track.

Defining Business Requirements

Chapter 7 will explain the need for the right approach in eBusiness requirement gathering, analysis, and documentation. During this process, the needs of the customer are gathered and translated into the specification of what the systems must do. There are seven major activities involved in this phase: business rules, content and data, application availability, user interface design, system integration, reporting, and security. This chapter will provide relevant approaches for

25

each of the areas and templates for the deliverables. This detailed approach will not only help to build a successful prototype for the project, but also lead to choosing a better architecture.

Designing an eBusiness System

Chapter 8 will examine the expected contents of an eBusiness System Design deliverable. The deliverable document from the system design process will include a set of technical specifications that will be used to build and implement the new system. A detailed System Design deliverable should address six major areas: the overview, application design, system design, technical design, next phase, and sign-off. An important aspect of this deliverable is that it should answer the "how" questions associated with a system, whereas the requirements phase deliverables should have answered the "what" questions.

Building and Implementing an eBusiness System

Chapter 9 will discuss the tasks involved with the building and implementing phase of an eBusiness project, which deals with the installation, customization, unit testing, and integration testing of the system (application) modules, screens, and reports. It will present the general order of installation and configuration tasks, how the tasks relate to each other, and where to find the information needed to complete them. The tasks involved will include: planning configuration, setting up databases, installing and configuring the eBusiness software components, configuring the existing systems to work with eBusiness systems, and testing. A detailed explanation of testing standards is provided at the end of the chapter for the QA part of an eBusiness project.

Securing the eBusiness Application

Chapter 10 will discuss the importance of securing an eBusiness application, law and Internet security issues, security advocacy organizations, types of security threats that an eBusiness can be exposed to, and security strategies and techniques that can be used to protect an eBusiness application from hackers. The content in this chapter assumes that the security of the eBusiness application will be designed, implemented, and maintained by individuals within the organization.

Protecting Consumer Privacy

Chapter 11 will set a baseline understanding of the online privacy landscape and how to lead your project team through the process of designing a privacy strategy that is effective, actionable and manageable. It is also designed to provide you with information on organizations that you can go to for support (e.g. keeping up with new privacy threats).

Deploying the eBusiness Project

To help an EPM organize the deployment strategy, Chapter 12 will define the stages of a roll-out strategy, discuss factors involved in developing a support strategy, and explore some methods and strategies for training end users.

Chapter 3

Selecting A Software Development Life Cycle

Chapter Preview

To be in a position to manage an e-project and guide a project team to a solution that meets the project objectives, an e-project manager must have a good understanding of the various software development life cycles. Without knowing the advantages and disadvantage of various software development life cycles, it would be difficult, for an e-project manager to manage the risks associated with various life cycle models and whether or not those models could deliver timely results.

In this chapter, seven of the most popular software development life cycle models used to deploy eBusiness solutions will be discussed. It will also explore the specific advantages and disadvantages of each model and how each model may be used in various applications to meet a project's objectives and goals. As you review each model, be sure to focus on the model itself, rather than on the method and unique application of a particular project, and how to best use these models to meet a project's objectives.

The key is to recognize which model may work best for a particular project objective. Because each e-project is unique, you must also have the ability to recommend variations of particular models to your project team. The last section of this chapter will give some guideline criteria and a sample evaluation matrix for determining the right model for a project.

Determining the Software Development Process

EBusiness requires continuous development to keep data fresh, the Web site "sticky," and ideas innovative. With a pure software development process and management strategy, we could easily fall into the trap of starting many projects and never completing them—or even worse—completing projects that are lacking in quality, over-budget, and late.

To avoid these situations, an eBusiness project manager should be knowledgeable of the various software development life cycles as displayed in Figure 3.1 below. Each model has specific advantages and disadvantages, but all can be used in various applications, depending upon the project's objectives and goals. An EPM may use any one or a combination of the various software development life cycle models available today to significantly reduce the amount of time and cost needed to produce valid results and achieve strategic business objectives.

Figure 3.1

Software Development Life Cycles

Model	Use	Pro	Con
Code and Fix	Smaller projects, short duration.	With user interaction, the work can be completed quickly.	Inexperienced developer can get trapped.
Waterfall	Project with well-defined requirements and delivery stages.	Easy-to-manage, step-by-step process.	Little or no opportunity to revisit previous stages for additional input.
Spiral	A risk-oriented model that breaks up a software project into smaller projects.	Less time developing the concept of operations, then developing requirements.	This model, if not carefully managed, can exceed time restrictions.
Evolutionary Prototype	Good model when requirements are changing rapidly.	Allows for refinement of code until the requirements are met.	Does not work well with time restraints.
Staged Delivery	Used for large development projects with well-defined requirements.	Easy to determine the status of the overall project and whether or not to continue.	Does not work well with poorly defined requirements.
Design to Schedule	Good model for "e" projects with specific and firm completion times.	Easy to manage risks.	Does not work well when the requirements are fuzzy.
Evolutionary Delivery	When a major feature to an existing program must be added by a certain date.	Provides good progress visibility to management and customer.	Developer must understand the architecture very well to be effective.

Take a look at each of these in more detail and the different applications for which they may be used.

Code and Fix

The code and fix model is just what it sounds like: a model in which code is developed and tried, then fixed until it is either accepted or meets the systems requirements as stated. Therefore, for a project where the requirements are not clear and the level of developer sophistication is not a critical factor, this model may be useful.

Again, there are drawbacks to consider when deciding upon any development model. As you progress in this model, code begins to get more and more complicated as the requirements either become clearer or more complex. This begins to impact time and cost as the project progresses and will tend not to allow for traditional scheduling. On the other hand, if time and cost are not the driving factors for a project, this model can be very useful.

For example, if an EPM was developing a new feature for an existing application that was relatively small with very clear requirements, and that feature had never been built before, the code and fix model might be a good method to use. It would allow him or her to develop the application and improve upon it as it evolves. Because it is conceptual, this would be a good model to get a product from idea to reality.

Pure Waterfall

A pure waterfall model is one in which the proceeding task must be completed before moving to the next. For instance, all of the coding must be completed before testing. Once a project begins with the use of this model, each of the respective modules must be completed to give the appropriate information or output to the next module, so that it may start. For example, the flow for a waterfall model may look like this: Concept → Requirements→ High-Level Design→ Detailed Design→ Coding→ System Testing→ User Acceptance Testing→Deployment→Post Implementation Support.

Many companies today claim that they are using this model. Most of the time, these instances are really variations of the waterfall model where other activities begin prior to completion of the previous activity. Not to imply that this cannot or should not happen, because some concurrent work can be done without jeopardizing the integrity of the project, as long as the work is not dependent upon or does not go beyond the preceding module's completion.

For instance, user acceptance tests should not be started while coding is still being completed. However, the development of test scripts could begin, based upon coding already completed, as long as it is understood and agreed upon that there may be changes required based upon the final code.

Figure 3.2

Waterfall Model

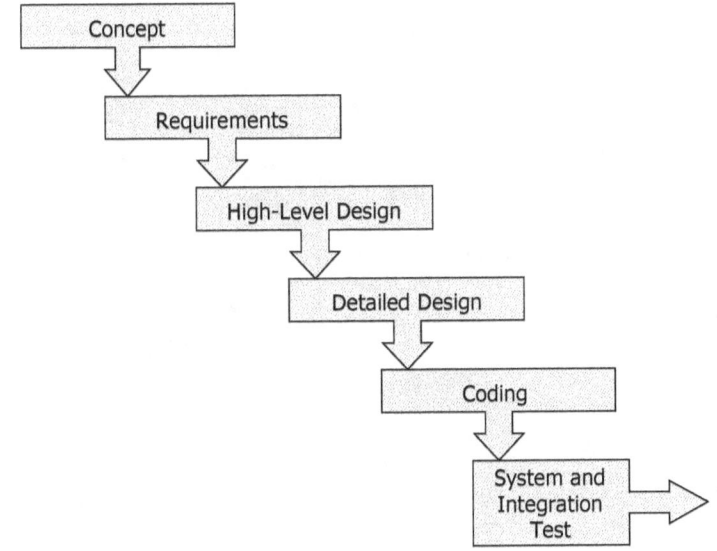

<u>*Spiral*</u>

The spiral life cycle is a risk-oriented model that breaks up a software development project into smaller projects. Each smaller project addresses one or more of the major risks, until all the major risks have been addressed. In the spiral model, the early iterations are the least expensive. Less time will be spent developing the concept of operations than the requirements.

The spiral model is a "best practice model," and if used, can prove to be very effective and produce a very reliable product. When using the spiral model, like in the previous model, the use of concurrent processing or engineering is very likely. Because the spiral model encourages revisiting and examining the previous work as the next iteration is begun, the project may seem like it is always taking a step in the wrong direction. With enough experience using this model, you may find this model to be the one most often used, because it encourages the team to review and confirm the previous work, while leaving open the possibility to revisit that work at any stage of the project as needed.

Figure 3.3

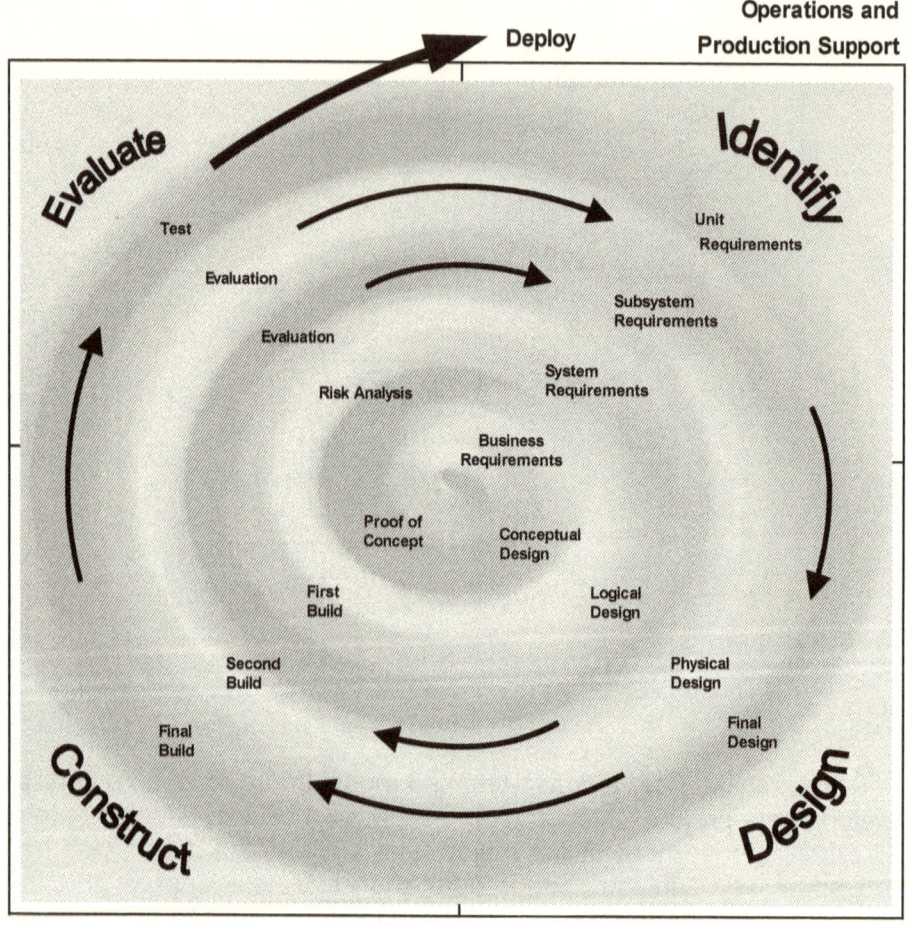

Evolutionary Prototype

The evolutionary prototype model is especially good when requirements change rapidly on a project, because this life cycle does allow for refinement of the code until the requirements are achieved.

This model applies where prototyping is needed and required. (Prior to the mass marketing of many products, a prototype is developed and used for testing and usability.) In figure 3.4, all of the iterations are in boxes except for the prototype work, which is in a circle, because this type of work can continue in a loop for long periods of time if not managed properly or given cost and schedule constraints. For this reason, this type of model does not work well with time restraints.

Figure 3.4

Evolutionary Prototype

Start by designing and implementing the most important modules of the project in a prototype, then add to and refine the prototype

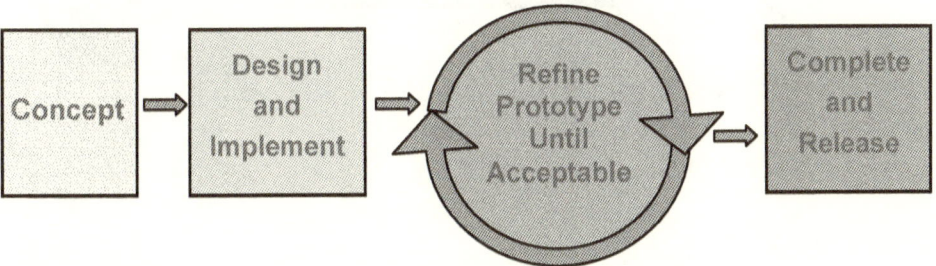

Staged Delivery

Staged delivery, also known as phased functionality, is used to deliver large development projects with well-defined requirements. This type of model can be used in fiscal planning, because it allows for budgeting for the project. From a customer or management view, this type of model allows for easy determination of the status of the overall project and whether or not to continue with the project at predetermined checkpoints. Using this model in large-scale implementations will also allow for more refinements or mid-course corrections as the project progresses, if needed.

An example of this option might concern project funding. This model would permit the financial sponsor to have the ability to stop the level of functionality at a given dollar amount without jeopardizing the entire project.

In this model, software is delivered in phases, and the requirements are well defined. Each phase is more refined with more features. The final product is not delivered at the end of the project. Project success is measured at each "staged" deliverable.

Staged Delivery

Figure 3.5

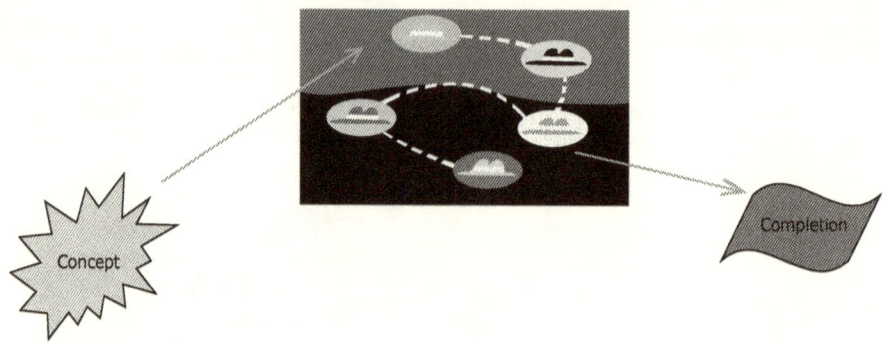

Design to Schedule

A design-to-schedule model is similar to the stage delivery model, except that it has a definite completion date. Using this model, the highest priority is addressed first, then the development moves to the next level of priority, and so on. The project is driven by time with the limited amount of available time allotted primarily for the higher priorities, so that they get the most focus. Inevitably however, the medium or lower priority requirements do not get addressed or the focus is not given to these requirements within the time needed, thus they get dropped. Although it's the "biggest bang for the buck" model, the project managers are the ones that will be there in the end with a net to catch the excluded functionality.

Figure 3.6

Design to Schedule

Evolutionary Delivery

Evolutionary delivery is a model where software is developed in stages or steps, where each step adds features, functionality, or value to the previous. A good use of this development effort would be for a concept or product that has not been previously developed and the concept has yet to be fully developed. This model provides the development team a tremendous amount of latitude to develop the product or concept with minimal constraints. Using the evolutionary delivery model, one straddles the boundary of evolutionary prototyping and staged delivery. A good application for this model is where the project size is anticipated to be large with a lot of growth.

Figure 3.7

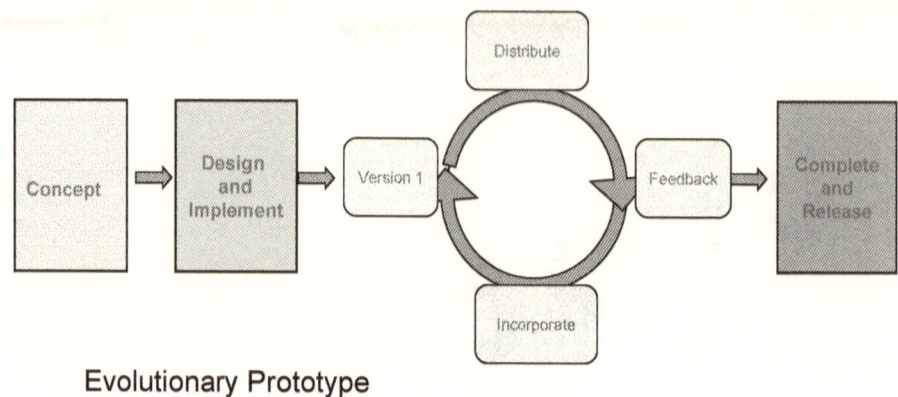

Evolutionary Prototype

<u>Selecting the right software development lifecycle model for your project</u>

How do you know which model to use for the project you are managing? Unfortunately, there is not one answer as to which model is best. When trying to decide which particular model may best map to your specific application, time line, budget, and resources, the following questions should be given consideration:

- How much reliability do we need?
- How much do we need to plan and design ahead for future projects?
- How well do we understand the system architecture?
- Will we need to make many mid-course corrections, and if so, what might the impact be?
- How well do we (or the customer) understand the requirements at the beginning of the project?
- How much risk does this project involve?
- What are the impacts if we fail or deliver late?
- Is this a mission-critical application?
- Are we constrained by funds, time, speed to market, competition, or internal pressure, like government requirements or mandates?
- How much of the project deliverables and progress will need to visible to the stakeholders or management?
- Do we have the resources to manage the project?

- If we do not have the resources to manage the project, do we know where to get them?

Would these be the only questions that would need answers? Probably not. Some other considerations might include experience using a particular model, experience with similar projects using a particular model, and lessons learned on other projects using that model. To comprehensively answer the questions that would help indicate which model is right for a particular application, an EPM should develop a matrix to evaluate and map out various options, then present or suggest these options to the team and executive sponsorship. The final decision should be a collective decision between the architect, development lead, project manager, and user group.

A sample evaluation matrix is provided in Figure 3.8. After reviewing the information on the various models, use this tool to help you define which model would work best on a particular project. Use this sample as a framework and insert the questions that apply to your specific needs.

Figure 3.8

Sample Evaluation Matrix

Ask each question and assign each model a rating. Total the ratings at the bottom of each model to determine your model selection.

5= Ideal 3= Acceptable 1= Unacceptable							
Project Questions	Code and Fix	Water fall	Spiral	Evolutionary Prototype	Staged Delivery	Design to Schedule	Evolutionary Delivery
How much risk if it is late?							
Mission-critical application?							
Constraints, funds, time, regulatory?							
How much visibility to stakeholders will be needed?							
How much reliability?							
Resources needed to complete?							
Understanding of the require-ments?							
Total							

For more information about defining software development, consult the following references:

1. The Software Engineering Institute is responsible for the Capability Maturity Model (CMM) and other valuable software development management processes. http://www.sei.cmu.edu.

2. The Software Engineering Laboratory at NASA's Goddard Space Flight Center is working on some interesting software development programs. http://sel.gsf.nasa.gov

3. The Software Productivity Consortium is a Department of Defense (DoD) sponsored consortium focused on software productivity improvement. http://www.software.org

Chapter Summary

This chapter defined seven of the most popular software development life cycles and also explored the various advantages and disadvantages of each model. It also provided some guideline criteria and a sample evaluation matrix for determining which model to use. For more information about these and other software life cycle models, you may want to consult one of the many books on the market, such as Rapid Development by Steven McConnell, Microsoft Press.

With a good foundation of life cycle models, you will find that many of the lessons learned in Chapters 1-3 will begin to crystallize as you learn how to define an e-strategy in Chapter 4.

Chapter 4

Defining e-Strategy

Chapter Preview

What is the strategy for eBusiness as we move into the new millennium? Will development, commerce, technology, or something else drive "e"? What is business is looking for, and how will project management play an essential role? Part of that answer is defined in the formula: change + speed = process improvement and efficiency. The strategy is to be the first to market, change the process and thinking, and be very efficient. Utilization and realization are terms that will evolve over the next few years as the Internet economy matures. Any successful business, of any size, will use these concepts to improve, reduce, increase, develop, deliver, and, most importantly, produce a profit.

One of the ways companies are using project management to help them achieve these results is to turn the company's business objectives into projects that can be delivered or converted into a product or service delivered on the Internet. Now that a foundation has been laid that defines the role of an EPM and various software development life cycles, this chapter will begin to explain how to strategically approach the management of e-enabled projects in order to realize the concepts above.

The development of an e-strategy begins by identifying business processes that can be e-enabled. During this process, an EPM must assess how the e-strategy will impact the organizational culture, productivity, and efficiency of an organization then determine the potential value contribution for an e-enabled application as it may apply before, during, and after deployment. Additionally, this chapter will also explore the factors to be considered when selecting an implementation model, provide a sample implementation table for identifying these factors, and discuss the option for an EPM to outsource or develop the applications in-house.

Identifying Business Processes to Be Enabled

When we look at the Internet today, we see many e-enabled applications that just a few years ago seemed outlandish to even consider. For example, there are Web sites that deliver convenience store items to your home in less than an hour. In this new economy of instant gratification and the increasing need for speed in everything we do, this type of an application seemed "natural." In some areas, this application has done quite well, but in other areas the company has opened and closed their doors with little fanfare.

Certainly, the idea of enabling business applications for the sake of being able to tout "eBusiness" in your company portfolio is not worth the investment, however many businesses today are realizing that well-planned and well-executed e-enabled business applications can prove to be efficient, cost-effective, and revenue-saving, as well as revenue-producing. One look at some high-profile, revenue-generating eBusiness applications, such as electronic bill paying, will quickly show that there are many ways in which a business can profit using this form of transaction exchange.

The role of the EPM is to be involved with the decision-making process of an eBusiness application from the evaluation stage to the delivery stage. Before making the decision to e-enable a business application, many factors must be considered. Knowing the what, why, and who can be the most important components of project success. To answer some of these questions, an EPM should assess the organizational impact of an eBusiness application and determine the potential value contribution of an eBusiness application.

Assessing Organizational Impact of an eBusiness Application

History tells us that change is constant. What is not clear is how organizations handle change and how change is managed. This unknown leads to an area of our strategy discussion that begs for attention. The impact on the culture of an organization when change is introduced is, at best, unpredictable. When developing an e-strategy, adding this element of risk into the equation is prudent thinking.

It is the EPM's charter to manage the project in a way to reduce the possible negative impact of critical areas of the project. Although these critical areas are discussed when consideration is given for any project, they become more visible in eBusiness projects. When assessing the organizational impact of an eBusiness application, you should examine the eBusiness application's potential impact upon:

- Culture
- Productivity
- Efficiency

When discussing culture, efficiency, and productivity, it is important to note that some of the changes in these areas have been very positive. A

41

report by Jupiter Media Metrix Inc. (Jupiter Research Unit), completed in January 2001, noted, "The number of employees with high-speed Internet access in the workplace is expected to more than double from 24 million last year to 55 million by 2005." As more and more high speed access is distributed and the costs drops, productivity and efficiency will increase. This cultural change will move workers and business closer to the realization that it is not the information age that is critical, it is the age of your information that should be managed.

Impact on Culture

With technology moving so fast and in so many directions, what are the impacts of eBusiness on the culture of both existing and new companies as they emerge onto the e-market space? Many companies are realizing that managing the cultural change within their organizations is in itself a challenge.

One objective of a quality management effort is to transform the culture of the organization into one that emphasizes a commitment to excellence; a focus on customers by everyone; and continuous improvement in processes, products, or services—a clear indicator that an organization is integrating quality management practices into its daily activities. The EPM should lead the effort to guide a project team with these principles, thus positively affecting the project and the organization as a whole.

Impact on Productivity

In developing an eBusiness solution, organizations often seek to improve their productivity and efficiency, both within the organization and within their external customer base and supply chain. Other principal drivers for companies to increase productivity is the need to remain competitive and reduce costs.

One of the newer entries in the software application arena is professional services automation (PSA). This software is not only a way to manage resources it also has more value-added propositions than may be apparent at first glance. Standardization of business processes and workflow management are two positive impacts from implementing a PSA tool that provide increasing productivity, especially for companies that sell services and resources. The EPMs, who realize that there is more to an e-strategy than just simply delivering a product or service, are the ones that will differentiate themselves. Understanding the business perspective of what you are implementing has great value to your organization or client. For example, when deploying a workforce management solution, one of the more important features a company will realize early on is that to implement this solution, business process must be agreed upon. This one important realization could be worth the price of the software even if it is never implemented. Think about having a standard business process across your organization and what that could mean in time savings, training, productivity, utilization and retention.

However, negative impacts on productivity must also be managed by the EPM. Such an example could include software management. Installing software updates can be a productivity drain. What is the value of updating to the latest version of a product on a regular basis? It can be difficult to measure these types of productivity aspects on a project.

Productivity, as it relates to eBusiness, is not as easy to define as it was in the Industrial Age. As an output in the Industrial Age, productivity was easier to measure. To measure productivity now, we tend to identify it by using more intangible terms, like innovative ideas, creativity, trust, and confidence. As the flow of information shows no sign of slowing, productivity can also be measure in time. That is to say, how much time does it take to complete various tasks, find information and then use that information for productive purposes. We realize a gain in productivity when we can reduce the time it takes to perform a task.

Impact on Efficiency

EBusiness project managers must ensure that a project will improve efficiency through such measures as an increase in revenues and profits, improvements in customer service, and a reduction in overhead and redundancy. However, an eBusiness project manager should achieve these objectives with more than just quality measurements. A successful application of quality management will consist of more than the application of selected tools and techniques or formal quality structures and procedures. Careful and collective management of an eBusiness project must include having the information and systems in place to make decisions at all levels on a daily basis. Advanced quality organizations have progressed to the stage where the application of quality principles and techniques are integrated to a considerable extent in the daily work of the entire organization. EBusiness project management is also about empowering the team and customers, while improving the bottom line by efficiently delivering quality products and services.

Each eBusiness solution is different for each company. While one company may need to provide online access to inventory for their sales team around the world, another company may need EDI capability with

43

their biggest clients while simultaneously maintaining their own proprietary system. Yet, another company may need both solutions integrated with a single solution for selling their products online. For the eBusiness project manager, the need to manage and employ business process efficiencies across the enterprise will be ongoing throughout the project.

Business process integrity, different than business process efficiency, is the design, implementation, and management of automated and process-based internal control systems. This is important to note because it drives the overall efficiency of the internal business process and how transactions are recorded, edited, and reported. You should have a basic understanding of these processes in order to intelligently contribute to eBusiness decisions made during the project.

Determining the Value Contribution of an eBusiness Application

In order to determine the potential value contribution of an eBusiness project, you must have a clear understanding of the business justification used to secure, justify, and validate funding for the project. Additionally, you must also establish criteria to evaluate success of the project as it develops. Defining critical and measurable success criteria before the project gets underway is a valuable exercise that will allow incremental success milestones to be defined throughout the project life cycle. This information is vital in making good decisions, recommendations, and adjustments to the project, thus facilitating a timely completion within budget and with the highest possible quality.

With all of the components described so far, including the unknown variables that will inevitably arise (deregulation, court rulings, market downturn, competitive pressure etc.), several questions must be answered to determine the value of the e-application, thus calculating the Return On the Investment (ROI). Many of these questions would ultimately be answered at the business requirements definition stage (this topic will be covered in more detail in Chapter 7), but before this stage, the should have solid answers to these questions:

- What additional value will this application bring to our business?
- What additional value would we be missing if we did not proceed with this application?
- How will this application map to our company's business objectives, both now and in the future?

- Can we do this any other way with the same or better results?
- Who will support this application after it is deployed?
- Will the success or failure of this application prove detrimental to any part of our business?
- Does anyone else have a similar application and what is their success rate?
- Is the timing right to launch this application?

Selecting an Implementation Model

The forces that drive an organization to develop an e-strategy are: a change in productivity, variable market trends, and competitive pressures. Marketplace, positioning, investment approach, value matrix, and migration path all play a role in developing an e-strategy approach. To avoid mistakes during a strategic implementation, you should identify all of these drivers. Many failures associated with e-project implementations can be traced to a lack of focus and coordination of these factors by the project team. An experienced EPM should be able to recognize these factors in advance and have the ability to take corrective action to reduce any chance of failure.

Understanding what drives these changes and why certain decisions are made is fundamental to project success. When selecting an implementation model for eBusiness, the existing (if not a start-up) sales and marketing strategy must first be examined or revised to focus on the business objectives as they will function in the e-environment. E-marketing enables companies to capture information about their customers. How they use that information is vital to the success and penetration in the market. When considering the marketing plan, you must consider all of the potential data mining that will take place, such as who will use the data, when will reports be generated, and who will see the reports and use the information to generate new business.

After the marketing strategy is examined or revised to focus on the business objectives as they will function in the e-environment, an EPM should focus on maintaining the integrity of the project objectives, while ensuring that the project objectives align with the business strategy throughout the duration of the project. Frequently, the coordination of the business and project objectives will usually include the long-term goals of facilitating marketplace differentiation and creating a "sticky" Web site.

To facilitate marketplace differentiation, how can a product, business, or Web site be made different than everybody else's? In eBusiness, that is what will separate the winners from those who are just hanging in there. Whether that differentiation is price, cost savings, first to market, or uniqueness, you must identify this differentiation during the determination of the marketing strategy and ensure that it is clearly communicated to the entire team.

Just as important, an e-marketing strategy will usually be designed to encourage repeat visits to a Web site, also known as making the site "sticky." Market research shows that it is more important to get and keep a customer than just building a large customer base.

As you can see, there are many factors to be considered when developing an implementation model. To help identify and assess these factors, Figure 4.1 displays a sample implementation table that may be used to help define your eStrategy.

Figure 4.1

Sample Implementation Table

Implementation	Currently e-Enabled	In Digital Format	Action / Issue
Marketing Plan			
Sales Strategy			
Manufacturing Process			
Product Delivery Methods			
Change Management			
Software Development Process			
Customer Support Plan			
Supply Chain Management			
Bill Payment Process			
Billing and Invoice Delivery			

e-Enabling Key Areas

After determining key areas to be implemented, the selection of the right application to e-enable each area is an important consideration, but it is only the first step. EBusiness presence will also, without question, demand support from a succession of critical technologies and suppliers. Without this support chain, you will default to providing the skills and devotion to your application when your organization may or may not be able to manage the workload. The EPM must recognize this and plan the project with the team to ensure that the burden of support is minimized at project conclusion.

To provide this support chain, you may consider what tasks will be handled in-house and what tasks may be delegated to outsourcing. Many companies are seeking to cut costs in the long term by outsourcing Internet functions to experts, such as desktop maintenance, data center management, and Web applications development. The most common areas that outside support is often needed include the development of new software; concurrent programming efforts to reduce time to market; support positions for project management; and individual specialized skills, like EDI (Electronic Data Interchange: a set of standards for exchanging orders and other business transactions by electronic mail) programming. Other areas where outside resources or agency resources are frequently used include video and audio production efforts.

It is now possible to outsource design and delivery of a complete e-enabled system. This offers a turnkey solution (supplied, installed, or purchased in a condition ready for immediate use) and isolates the costs and results from other activities. A turnkey solution is often the only way to install a major new system without hiring a massive number of new personnel. It offers the advantages of separating objective setting, performance measurement, and the delivery of a system. If managed well, it forces the clarification of the scope of work. (After all, EPMs should never enter into a binding contract for a system that does not have adequately defined, clearly stated, and obtainable specifications.) If managed poorly, it will create legal disputes that will force an indefinite delay of the new system.

Failed turnkey contracts in the private sector remain hidden from public view, because everyone is interested in forgetting the fiasco as quickly as possible. A U.S. government agency, however, cannot always hide its aborted systems contracts, and therefore offers a rare opportunity to understand how to manage turnkey contracts.

For example, as a result of delays in the U.S. Patent Office's information systems overhaul, it was decided that a turnkey vendor, who could rapidly deliver the system, would be used. The new system would place all of the Patent Office documents "online" for search and retrieval. Because of the urgency and uniqueness of the task, the Patent Office obtained a waiver so that it did not have to follow established Federal vendor selection, vendor evaluation, and competitive bidding procedures. The Patent Office proceeded to select one vendor who seemed to be the most responsive to its need for a rapid product delivery.

47

Shortly after the contract was awarded, it was discovered that improvements in searching for and retrieving documents did not create budget savings. As a result, the Patent Office decided to change the scope of the project. The emphasis shifted from clerical efficiency to improving the quality of the patent search by the patent examiners. The contractor then spent 18 months negotiating changes in contract terms and systems specifications. Meanwhile, money continued to flow for software, for which there was no agreement and no hardware yet available, because of a lack of supplemental funding. Four years and $448 million later, it is still unclear what benefits the Patent Office will get from its new computer system.

Although outsourcing has its pros and cons, it doesn't mean that it should not be correctly managed. This is even more relevant for the e-enabled application. The EPM will be the person on point to manage the overall progress and compliance of the project to the contract standards and specifications. Prior to any contract award, the desired end results must be unambiguously spelled out. Upper management must also limit unavoidable minor modifications in the original systems specifications from accumulating into a different design. Commonly known as scope creep, this is more profound on eBusiness projects, as change is constant within and around the project. When an implementation is handed over to the EPM, the project will be a disaster if everyone tries to accommodate changes without thinking about their cumulative consequences.

When considering outsourcing options, an EPM may correspond with various sizes of outsourcing companies. With budgets for eBusiness solutions being expected to increase as much as 30 percent and with the market having been as rough as it was during the first part of 2001, many companies are now looking at ways to accelerate their long-term outsourcing plans. The benefits of this type of movement will be realized by the larger, more established companies with a proven track record, who back up what they say they can do through continued customer retention and growth. This is not to say that small outsourcing companies are unstable. Smaller companies have their advantages too, as long as they are able to back up those claims with solid references, value-added methodologies, and proven skill sets for what the EPM's company is looking for.

Conversely, if you decide to have particular tasks e-enabled through in-house resources, you should have already assessed organizational impact, but would still need to give thorough consideration to the type of application to be used. There are products and services that have proven themselves over the past few years to provide the passion and devotion to the components that will promote success. The use of products that can facilitate and assist in the tasks required is acceptable and encouraged. Although some of the applications may be substituted or omitted, depending upon the type of project you are working on, Figure 4.2 lists some of the products that may be considered.

Figure 4.2

Product Suggestions

Product / Use	Product Name	Supplier
Security: Intrusion Defense and Response	PitBull LX for Lunix	Argus Systems
Integrated Programming Tool Sets	Web Gain Studio	Web Gain Inc.
Internet-Based Multimedia	Live Motion	Adobe Systems
Customer Relationship Management	e-point 5	Point Information Systems Inc.
Developer Productivity	Rational Suite Enterprise	Rational Software Corp.
B2B Platform and Software Services	Business Communications Manager	Nortel Networks
Financial and Accounting	Eenterprise 6.0	Great Plains Software Inc.
e-Commerce Implementation	Commerce Server 2000	Microsoft Corp.

Chapter Summary

This chapter discussed the processes of identifying business applications to be e-enabled, assessing of the potential organizational impact of e-enabling such applications, and estimating of the potential value contribution that the considered applications may provide. It also explored some of the factors that you must consider when selecting an implementation model and provided a sample implementation table for assistance with identifying any additional factors that may apply to individual projects. The pros and cons of outsourcing vs. developing an application in-house were also discussed.

With the knowledge of the steps in involved with developing an e-strategy, you are now ready to move forward to Chapter 5, where you will learn about the planning stages of an e-project.

Chapter 5

Planning the Project

Chapter Preview

With so many different methodologies available to guide an EPM through the planning phase of project management, it can be hard to decide which one to use.   When you are engaged in an eBusiness implementation, the differences (as they pertain to e-projects) can become very blurred.   However, if you already have a background in project management, this is not the time to discount your past experience. When planning for an eBusiness project, there are a variety of topics that you must be familiar with to succeed and calling upon that past experience may be the difference between success or failure of a project.

To begin to clarify these issues, as they apply directly to the planning phase of an eBusiness project, this chapter will discuss many of the different aspects of planning an eBusiness project, such as the steps involved in managing risk; constructing a project charter; selecting and working with vendors; funding an eBusiness project, including identifying, estimating, and controlling costs; and developing a Work Breakdown Structure.

Risk Management for the EPM

Risk and uncertainty always accompany the decision to proceed with a project, because the project activities and outcomes are planned for the future.   The sheer duration and complexity of most projects virtually assure that problems will occur and need to be managed.   However, distinct differences in risk management exist for traditional project managers and eBusiness project managers.   While traditional project managers would define risk as specific, an EPM's definition of risk should be more intrinsic, thus an EPM should have more of a "risk-acceptance" view.

As the project manager, you are responsible for all aspects of the project, including those that are largely out of your control.   Going into a project with a "can-do" attitude may support team spirit at the onset, but when the project begins to fall apart and reality sets in, the team will become frustrated, the customer will be unhappy, and the EPM will regret not having taken a hard and realistic look ahead at what could go wrong.

While there are no guarantees regarding the future, the EPM can begin to identify and mitigate risks, as identified by the team or others, by taking certain steps. The four steps for managing risk include:

1) Identifying areas of significant risk
2) Performing a risk assessment
3) Designing actions/strategies to mitigate risks
4) Communicating significant risks to the key stakeholders

Risk areas that an EPM should pay particular attention to are schedule slippage, budget overruns, quality compromises, and dependencies on suppliers to deliver, such as outsourcing software development. An EPM must also consider all of the obvious risks that are valid with any business, like competitive pressures, regulatory or legal issues, and whether or not the technology is compatible with the target market or company growth. Testing is another area that should be examined, with the traditional view in mind of achieving predefined quality targets. The EPM should be geared to do this, to a lesser extent, and be focused on controlling and managing risks that could result from testing.

To help minimize other specific risks, an EPM should never:

- Commit both to budget and schedule for an application that has never been delivered on budget or on schedule.
- Base the implementation schedules on vendors' promised delivery dates for software.
- Convert an old application to a new one without a confirmed back-out plan with retraceable steps to restore the old system.
- Program an application in a programming language that is known only to a small percent of the staff.
- Allow modifications to the vendor's operating system.
- Rely on 100 percent availability of a single communications link.
- Assume responsibility for running an undocumented system.
- Contract the development of an application on a time and material basis.
- Start up a system without a prior acceptance test from the key stakeholder, client, or end-user.
- Introduce a mission-critical application using totally new technology.

Although each e-project is unique and has its own character in many ways, each is also similar in the sense that the EPM must be the one to consider all of the possibilities and contingencies and plan for them to the last detail, then the project manager must tell *everyone* involved—then tell them again. The EPM must realize that nearly everyone in the organization will be affected, such as shipping, marketing, legal, communications, human resources, and recruiting. All of these functional areas, and more, would no doubt be impacted in some way.

Planning the Project

Regardless of project type, scope, or duration, there are certain basic issues that must be addressed when considering deployment of an e-project. As with all projects, eBusiness projects will be most successful if these elements are managed in a proactive fashion. The topics of schedule management and real-time deliverables, as they apply to the EPM, could consume half of this book, but certainly worth mentioning is the fact that the EPM must be able to carefully calculate and deliver a fine balance between product and schedule, quality and cost, and do it all within an accelerated delivery window driven by market opportunity.

The speed-to-market global adoption of a product or service is usually of primary importance. The process of release management (in this usage, managing ongoing software development and deployment), of a traditional product release was meticulous and detailed. The EPM release strategy should be focused on speed-to-market and RAD (Rapid Application Development) of the new product. In order to stay competitive, Web sites must continuously produce fresh content to keep the company's sites sticky.

To accomplish these goals, one of the first project documents that must be completed when the initial planning begins is a project charter or project initiation plan. EBusiness projects have a tendency to grow as they mature. When bringing an idea to reality, it will usually take on a different look than the original vision. The project charter will help to measure success by providing a firm baseline measurement to review at any given point in the project. The project charter is the one document that you should refer to throughout the entire project, as a reminder of the project objectives, what was agreed to, and by whom.

One of the outputs from this plan should be a smaller, high-level plan that identifies the high-level milestones that need to be reached. The intent is to capture the larger tasks that need to be completed on the project and identify deliverables (or output) and owners of those tasks.

This high-level view can then be used as the foundation data for a Work Breakdown Structure (WBS). A WBS is a tool for defining the hierarchical breakdown and work in a project. It is developed by identifying the highest level of work in the project. These major categories are broken down into smaller components. For example, the highest level of work for a Web site development project would be the project itself followed by the next level which would be the hardware, software and the product service. Continuing to breakdown the work into smaller packages as in the example below.

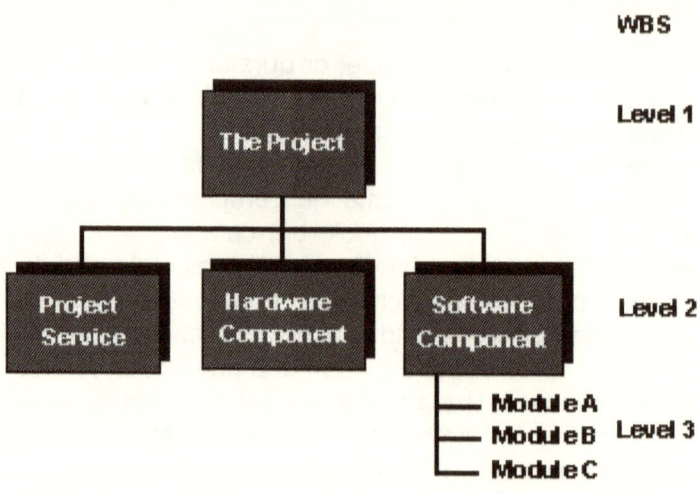

As a sample approach that can be used for planning and executing e-commerce projects, Figure 5.1 displays an example of a high-level plan showing what information needs to be captured, descriptions of some of the major stages, and a detailed list of tasks that can be used on e-projects.

The primary purpose for including this information is to provide a context from which eBusiness projects can be delivered or developed. It is not meant to replace the existing Systems Development Life Cycle, nor is it intended to represent the only way to plan and schedule eBusiness projects. It is simply an alternative that works, and one that can be used to position issues that will be faced on eBusiness projects.

Figure 5.1

Sample High-Level Plan

Task	Task Description	Days to Complete	Estimated Cost	Deliverable	Owner
1.	Manage Project	300	180,000	Various	Project Manager
2.	Define Requirements	100	60,000	Requirements Document Completed	Project Manager, Requirements Team
3.	System Investigation	50	30,000	Capability Determined	Lead Architect, Lead Technical Developer
4.	Media Integration Strategy	75	45,000	Design Document with Cost Matrix	User Experience Group
5.	Content Management Strategy	100	60,000	Content Collection Plan	Marketing
6.	User Testing	40	40,000	UAT Test Plan	Help Desk
7.	Post-Implementation Support Plan	365	300,000	Support plan	Technical Resources
	TOTAL	1030	715,000		

Selecting and Working with Support Vendors

How will you get all the work that is forecasted in a project charter completed if only limited resources are available? One solution is by selecting and managing agency resources that augment the existing resources. It usually sounds easier than it is to do, considering the need to determine the specific skills required to complete the project and the many choices that are available to do so.

Would you select a contractor to paint your house if you had never seen their work? Of course not. Neither would you select a contractor to outsource a Web development project without seeing their work. Without knowing the level at which a contractor, supplier, or outsourcing company will perform, an EPM cannot ensure that the project expectations will be met.

One method used to validate potential suppliers is the use of a Request For Proposal (RFP), which defines, in detail, the scope of work required and the expected level of quality that is to be delivered. The RFP should always be developed with the acceptance criteria predefined so that when the responses are received, they can be measured against how well they map back to the success criteria. This takes any emotion out of the selection process, while providing a clear and repeatable process methodology for future outsourced work, from which responses can be assessed.

Another method for selecting vendors, or a complimentary method to the RFP, is the use of a selection matrix, which can define and document the specific criteria that will be needed from the supplier. This type of selection process can be used over and over again and modified to meet specific needs for different types of projects. The Supplier Selection Matrix in Figure 5.2 can help to distinguish the "needs" and "wants" for selection and determine as to whether or not the proposed suppliers meet all of your selection criteria. Start by defining what the decision factors will be, in other words what are the most important considerations to base a decision. Enter that information in the Decision Factor column. Next take your decision factors and divide them into "Needs and Wants" and enter that information in the appropriate box. Add weighting to the items in the "Wants" box. To do that select the highest weighted "Want" and then move to the lowest, keeping in mind that your weights should be equal to one. Weights are used to decide how much emphasis should be placed on each of the factors (Wgt. In Wants). Enter the name of the two or three selected suppliers next to the "Go" box. Now moving left to right, start with your "needs" and decide if each supplier is a "Go" or "No Go". Once that is done and if you have not checked any "No Go' boxes, then move down to the "Wants" box. Look at each "Want" and score them by assigning a number from one to ten. After scoring each of the needs and wants, then assigning a weight to each, multiply the weight times the score to give a weighted score (WT, SC). Finally, total the columns to see which supplier meets the selection criteria the best.

Figure 5.2

Supplier Selection Matrix

Decision Goal:		Decision Timing: ❏ Is there a date by which this decision must be made? _____ ❏ If so, what happens after that date? _____ ❏ If no, what time frame is realistic? _____									
Decision Factors:	Needs: (Go/No-Go Measurement, Essential, Realistic)	Alternatives									
			GO	NOGO		GO	NOGO		GO	NOGO
	Wants:	Wgt		SC	WTSC		SC	WTSC		SC	WTSC
			Total			Total			Total	

After a qualified vendor has been carefully chosen, working with one can still be a challenge. The paradigm of the new economy is stretching IT budgets and staff resources like never before. With the eBusiness model encompassing the complete spectrum of infrastructure components, such as servers, networking, storage, performance management, security, and much more, building and managing the scope and complexity of this kind of an enterprise-scale infrastructure is something that few organizations have the capital, time, or expertise to handle.

However, there are alternatives to the do-it-yourself approach of managing resources. A wide range of service providers of different types and capabilities can offer assistance with the arduous task of moving from traditional business to eBusiness. These companies can step in and take charge of some or all of your IT (Information Technology) and network infrastructure needs. Some of the companies available today can offer services as wide as a complete end-to-end solution, including hosting and post-production support. Other companies can offer boutique services, such as specialized programming, content management and development, or simply project management support.

By focusing on its business of managing the infrastructure that underlies its clients' applications, an outsourcer can free a company's in-house staff to stay focused on core business issues and bottom-line goals. If performed well, outsourcing can deliver an improved application uptime, which will translate to revenue growth and greater customer loyalty and retention—all for a controlled, predictable fee.

However, buyer beware: Not all outsourcing is created equal. Many first-generation hosting providers offer a piecemeal infrastructure comprised of technology components that cannot keep pace with the demands of eBusiness. Typically, their customers are left with the burdens of not only application development, but also integration, implementation, and equipment provisioning. You should be selective and do his or her research *before* deciding upon an outsourcing partner. The term partner is chosen because that is exactly what the relationship between your company and the vendor should be: a partnership between your company and the vendor to complete the project as a team. The vendor must have a stake in the project. If the EPM's company does not have a contract management organization, then the EPM must make every effort to ensure that the selected supplier is going to deliver what is expected, when it is expected.

In today's marketplace, across all industry segments, businesses are realizing that the transformation to eBusiness is required to remain competitive. Analysts predict that the companies who are not making the necessary changes will be overrun by competition and could ultimately fail. As enterprises around the world undergo this transformation, they are increasingly leveraging outsourcing options to help:

- Broaden their markets by extending their reach globally
- Enter new business areas through collaborations or expanded services made possible with Web-based interactions
- Increase employee productivity by providing easier access to corporate information and services
- Reduce costs through improved operations that integrate Web-access and traditional Information Technology (IT) systems

Funding the eBusiness Project

Identifying and Estimating Cost Elements

Until the project scope and plan has been completed, you should not try to estimate the cost of the project. Unless asked to give a range estimate or an order of magnitude estimate, there will not be enough information to venture a guess.

Many unknowns are inherent in eBusiness when it comes to estimating and budgeting costs. The primary reason for this is that many of the components of the project will not have been completed before and must be developed, either in-house or outsourced. In order to capture costs and develop estimates, you must be able to identify as much as possible about the level of magnitude that each element defined in the project charter will cost.

When estimating a project cost, there are several methods that can be used. However, the amount of information already available will have an impact upon which type of method will be used. For example, if a project has been defined well, with the tasks broken down to the lowest levels, the estimates of cost can be derived from those tasks, especially by taking high-level estimates to the respective owners responsible for delivery of the project and getting estimates from their perspective. This is called "bottom-up estimating." All of the tasks' costs are then summarized to give the overall estimate. Of course, the accuracy of the estimate will depend upon the accuracy of the data collected.

Another method of cost estimation is analogous estimating, which is a "top-down" estimating format, where previous projects (similar to the project proposed) are used to estimate the costs for the current project. You can evaluate previous cost estimates, lessons learned, or other project documentation to help with costing out the current project.

Preparing the Project Budget

Working with the project team to develop a project budget can be overwhelming. The amount of detail needed to support the budget can be a project in itself. Project budgets will require input from several sources, so understanding where to get that information is key. Many eBusiness projects will complete a cost-benefit analysis on the project before approval is given.

You will need to participate in this process to get a sense of what the stakeholders are truly expecting in the form of ROI (Return on Investment) on the project. When evaluating a budgetary number for project management and administrative cost, anywhere from 2 to 12 percent of the overall project cost should be considered. Depending upon the complexity of the work, the amount of new technology and past experience should be taken into account.

Controlling Costs

Of the many other duties of the eBusiness project manager, the most consuming is cost control. Controlling cost is really a balancing act between two other aspects of any project: resources and quality. When one of these is out of balance, it is due to either one or both of the remaining aspects being out of balance. For instance, as scope begins to creep on a project, costs often do so as well. If the schedule begins to improve, chances are that the quality is diminishing or the scope has been reduced. When considering how to control costs on a project, you must consider how to control other aspects of the project as well.

Scope creep, feature creep, and change control all play important factors in controlling costs. All of these issues must be covered in the project initiation phase and discussed, defined, and resolved long before the project gets under way. Another method of cost control can be attributed to a project charter that clearly defines success factors, quality expectations, and baseline measurements that clarify where the project started and where it should be when it is complete. A solid project charter or initiation plan, supported by executive sponsorship, will prove to be invaluable to the overall success of project cost control.

Continuous monitoring of the project time expended by the team, contractors, or outsourced work, through whatever accounting program available, will allow the project manager to forecast and manage costs. Different for the eBusiness project manager is the amount and type of personnel that will be charging time and efforts to the project, like concurrent contracting work, either in-house or outsourced, such as graphics and video production; Web Application Program Interface (API), or content development; and marketing efforts to assist in higher adoption rates and branding efforts.

Developing the WBS

Determining project steps and activities is a process by which project objectives can be achieved. This process will typically involve carrying out a number of tasks and producing a number of products during the course of the project. The tasks produce the products. For control and clarity of purpose, it is useful to arrange these tasks in a top-down structure, which progressively specifies the required work in greater detail.

It is important to look for opportunities to customize the WBS for e-projects due to the specific objectives of individual projects. The Work Breakdown Structure will provide a benchmark by which the quality of the project's process can be assessed.

While the project initiation plan defines what resources and associated time commitments are required to carry out the project, the Work Breakdown Structure provides a basis from which this estimate can be carried out. The resource and time commitment can be used to calculate an end date for the project and an estimate of its cost.

Activity Sequencing and Estimating Activity Duration

Activity sequencing is one of the first major tasks that will take place after the project has been funded and an initial plan has been finalized. The activity schedule will consist of all the activities identified in the high-level plan in the project initiation phase. These activities will then be decomposed into smaller tasks. The activity schedule reflects all of those tasks and the sequence in which they will occur. This schedule will also be used as means to track the overall project work.

The use of an activity schedule is probably the most effective way to determine how best to integrate available resources on a project. Activity schedules or sequences can be developed for the various audiences on a project. For example, a higher-level schedule may be used for management and project leads with a more detailed schedule for those who are performing the coding, testing, or other detailed level tasks.

When each component or task has been broken down to the lowest level, one approach to estimating activity duration may look like this. The responsible manager of a technical resource has been identified to perform a task, now they should submit a time estimate for that specific task. When they submit their estimate to the EPM, it should be in the following format:

- The most optimistic time to complete
- The most pessimistic time to complete
- The most likely time to complete

In order to combine these estimates into a single estimate for expected time, you can use the following formula in Figure 5.3, where T= Expected Time, a= Optimistic Time, b= Most Pessimistic Time, and M= Most Likely Time:

For example if, a= 3 weeks, b= 9 weeks, and M= 6 weeks, T would then equal 3+24+9=36/6 = 6 weeks. While understanding that this is a very simple example, take into account the delta between the optimistic and pessimistic, then decide if the estimate looks correct.

Another time estimation technique is the "expert-opinion" technique, based upon past experience with similar work. If a developer has experience with the particular programming language that will be used and a history with the type of work that is to be completed on this project, then this second method of estimation is acceptable, if it is considered as a range of time to completion. You should use this method sparingly or when no other data is available, as each perspective is different.

Chapter Summary

With the knowledge of the factors involved in managing risk, constructing a project charter, selecting and working with vendors, funding an eBusiness project, and developing a Work Breakdown Structure, you are ready to proceed to Chapter 6, where the basis of a project plan will be further defined with a discussion about managing team personnel and project stakeholders.

Chapter 6

Managing Team Personnel and Project Stakeholders

Chapter Preview

A single person does not win a baseball game—an entire team does. The selection and management of the core project team can be one of the most important actions, especially for an EPM, in project management. Leading the team in pursuit of mutual goals is a key part of the role the EPM plays. This is even more important on an e-project, because there is so many more opportunities for team members to feel excluded from the overall effort and viewed as simply a resource.

This chapter will discuss the areas of assembling an eBusiness project team, setting up a team architecture, identifying and documenting key skill requirements, identifying and communicating with project stakeholders, and keeping the team personnel, thus the project, on track.

Assembling an eBusiness Project Team

After developing the project plan and high-level plan, you should begin to identify/notify team members that will be involved in the project. Selecting and retaining the right people for any project improves the chances of a successful outcome. Of course, they all must have the correct skill set required for the project. Understanding and executing this point is key to any implementation, but even more relevant on an eBusiness project. Because many tasks are very compressed, there is very little opportunity to make up for errors in team selection.

If an EPM is given all of the resources needed, but they are not the right resources, the EPM is in just as much trouble as if he or she had none. Just as a homebuilder will need many resources to build a house, the house does not get built by having all plumbers as resources. Likewise, having fifteen really good JAVA developers on the team does not mean that the project will be completed any faster than with ten developers who work together routinely and just completed a project similar to the proposed one.

Another essential of project management is assuring successful individual and team performance on the project. The "right people" means more than just talented individuals; the individuals must work well together as a team. Candidates should certainly have the ability to change and respond to the ever-changing needs of the business, but most importantly, they should be team-oriented, receptive to working *with* the team, not just *for* the team. Just as important, they should be trustworthy. Trustworthy in that if they say they will be done on time, they will finish on time. Developing strategies to facilitate strong individual and team performance will greatly increase project success.

Setting Up the Project Team Architecture

The next step in planning an eBusiness project is setting up the project team architecture. An example of what the project team may look like is given in Figure 6.1. (A project may or may not include all of these team members, depending upon the type of eBusiness implementation.)

Figure 6.1

- Project Board
- Stakeholders and Users
- Project Manager
- Project Team
- Technical Lead
- Business Analyst
- Training Lead
- Test Lead

Sample Team Architecture

The Project Board, generally in the form of a Steering Committee, to review requested changes to the project and participate in the review sessions.

Stakeholders and Users to provide input via (Joint Application Development) JAD sessions and other interactive or iterative processes that define the project scope.

The Project Manager to develop the project statement, orchestrate the concept, manage the processes, and lead and guide the team so they do not lose focus of the business objectives.

The Technical Lead to lead the technical development team, provide guidance on infrastructure, and ensure that the development effort is cohesive.

The Business Analyst to work with the subject matter experts, understand the complete set of business requirements, and ensure that implementation meets functional requirements.

The Training Lead to incorporate product-specific material into the training documentation and organize training logistics, material, and delivery of the training program.

A Test Lead to organize and coordinate the efforts of the test team. The test team may or may not be internal and may be some combination of both. This role is essential in ensuring that your testing efforts remain focused, organized and successful.

To organize the team architecture and hierarchy, an EPM should establish a project organizational chart. An example is given in Figure 6.2.

Figure 6.2

Sample Project Organizational Chart

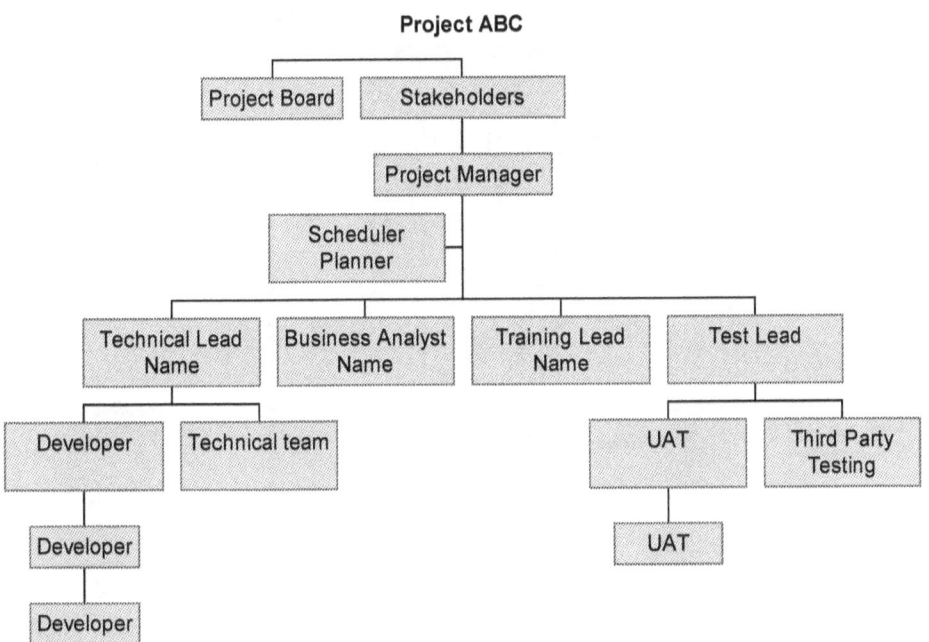

Identifying and Documenting Key Skill Requirements

After the project organizational chart is completed, the task of identifying key skills and requirements begins. The EPM should start by taking each of the roles identified to complete a task and then define each task in greater detail. This information should then be applied to a roles and responsibilities matrix to clarify the skills needed for each role. An example is given in Figure 6.3. It can be expanded to accommodate any project size. Additional columns can be added to meet specific needs.

Figure 6.3

Sample Roles and Responsibilities Matrix

For each activity, assign one or more of the following letters to complete the matrix:

A=Approve	C=Create	I=Inform	O=Optional
P=Participate	R=Responsible	S=Support.	

Activity	Internal Resources					Agency Resources		
Planning (Weeks 1-4)	Project Board	Project Manager	Technical Lead	Business Analysis	Training	Development	Testing
Project Kick-Off Meeting								
Define Project Charter								
Create Schedule								
Create Project Plan								
Develop Communications Plan								
Define Quality Plan								

Identifying Project Stakeholders

Stakeholders come in many different forms. Some may say that external clients are stakeholders, others may identify internal customers as stakeholders, and still others may identify a completely different group of people as the stakeholders on a given project. As the EPM, identifying with all of the stakeholders will be a key to project success. Not meeting the needs of just one influential stakeholder can create major problems for the project. Specifically for e-projects, understanding your target audience is very important. If you do not know how to influence that audience you may offend them, insult them or frustrate them.

When identifying project stakeholders, finding the "owner" of a responsibility and finding the one person who has signature authority is many times not easy or clear, because responsibility does not always mean authority. One of the EPM's initial functions is to identify who will ultimately "sign off" on a task or area of responsibility.

As the project charter is developed, identifying who will be responsible to authorize time, cost, and quality changes on the project is critical to success. To maintain the balance of the triple constraint (time, cost, and quality), the EPM must identify which stakeholders may or may not be willing to compromise one or any of those constraints. The take away for the EPM is to make sure they know who "owns" functionality and who is "responsible" for functionality.

Communicating with Project Stakeholders

Communication, or lack thereof, is at the heart of many project successes or failures. Communicating on an e-project is an art. Acquiring fundamental communication skills in project management will help the EPM to manage project expectation in a more effective way. The more proficient in communications an EPM becomes, the smoother a project will flow from beginning to end. How information is organized, maintained, and disseminated will have a significant impact upon the success of the implementation. It's a challenge to provide enough information to keep everyone on track and moving in the same direction, without overloading or boring them, but eBusiness projects have an inherent need for clear communication, often to a wide audience. Finding that balance is critical.

All projects face differences in what the client or end user wants and what ultimately is delivered. To avoid these types of issues, the EPM must know how to translate the requirements (or wishes) of the

67

stakeholder into a process or feature that can be measured as part of the project's success. By taking this approach to stakeholder requirements, the risk of not meeting the expectations of the stakeholder diminishes considerably. Having requirements, as measurable components, facilitates many other phases of the project.

The communications methods to be used to keep project stakeholders informed must be determined in a project communications plan. The communications plan should be part of the project charter and included as part of the charter documentation. The regularity of project communications, meetings, and project status reports must be decided and documented in this plan. Project complexity will often drive the frequency of these communications. Whenever possible, the stakeholders should participate in identifying these measurable results, which should become part of the success criteria when final acceptance and project closure occurs.

A sample communication matrix, identifying the source, frequency, delivery method, and audience, should be part of the communications plan. An example of a communication matrix is shown in Figure 6.4.

Figure 6.4

Sample Communication Matrix

Communication Description	Audience	Frequency	Medium	Person Responsible
Status Reports	Project Team Exec. Sponsor	Weekly, on Fridays	MS Word status report e-mailed to recipients	Project Manager
Project Status Meetings	Project Manager Project Team	Weekly	Status updates from Project Team members	Project Manager
Project Reviews	Project Manager Executive Team	End of each stage, based on project needs	Formal review of project status	Program Manager
Issue Discussion and Resolution Meetings	Project Manager Issue Owners	Daily, as needed	Informal discussions; e-mail exchanges	Project Manager
Project Schedule Updates	Project Team	Weekly, on Fridays	MS Project	Project Manager
Company-Wide Updates	All employees	Weekly	Intranet	Project Managers

Additionally, a marketing communications plan should be developed by the marketing team to ensure that the product or service has the best chance of success at launch. The marketing and communications plan is a formal plan that seeks to achieve and maintain active support of a product or service, using traditional marketing and communication techniques. A strategic marketing and communications plan, focused on building awareness and communicating the impact and benefits of the product or service, is critical to ensure success. Topics that should be included in the marketing plan are:

- Overview
- Target Audiences
- Marketing and Communications Objectives
- Marketing and Communications Strategy

For an internal project or a client-facing project, an example of a marketing communications plan can be seen in Figure 6.5.

Figure 6.5

Sample Marketing Communication Plan

Task	Communications Objectives	Audience	Resources	Date
Develop Marketing and Communication Plan	Establish communications strategy, objectives and timetable	Project Team	Project Managers with assistance from Marketing	March 27
Develop Feedback Mechanism	Create interactive two-way dialogue	Project Team	Discuss with Project Team	March 27
Write Communication #1 (to target audience)	Project objectives, process, impact on the organization, time frames	Entire Company	Project Managers with assistance from Marketing	April 15
Write Communication #2 (to target audience)	Update on project objectives, process, impact, time frames	Entire Company	Project Managers with assistance from Marketing	May 15
Evaluation	Monitor feedback, revise strategies if needed	Project Team	Discuss with Project Team	June 15
Produce Collateral/ Case Study	Provide educational workshops and develop other communications	Entire Company	Project Team input and approval	July 30
Road Show	Provide briefings, share lessons learned, focus on success	Entire Company	Work with Project Team to determine forums and timing	June/July
Final Evaluation	Review entire communications strategy against objectives with team. Document any gaps, successes for future plans	Project Team	Provide template for Project Team input and approval	June

Keeping Team Personnel (and the Project) on Track

One of the hardest issues facing companies today is getting a product to market in a timely manner. Idea-to-market turnaround time has decreased over the past few years. A product or service has to be delivered quickly if it is to be successful, especially when the product or service is Internet hardware or software related. It has been said (by Internet research companies and published in CIO magazine) that if an Internet-based product or service cannot be introduced within two weeks of the competition, over 50 percent of the market share has been lost. Conversely, if the product or service is the first to market, that market share has been gained, along with a competitive edge over the competition. First to market plays a big role in eBusiness success. What role, then, does the project manager play in the success of this effort?

The EPM is the one person who is able to look across the company's organizational components and rally the various executive sponsors to focus on the task at hand. Not only is there a need for this coordination, but also the need to manage to a schedule and keep the entire team focused and motivated to meet or exceed deadlines. Without focused guidance, communication, and management of the project, many talented people will all be working within their various areas of expertise, but not moving in the same direction with the focus to all finish at the same time.

As an example, could a house be built faster if simply more resources or people and equipment were added? The answer is yes and no. Simply adding more people and equipment won't get the house built any faster if they are all carpenters. To accelerate the process, the exact mixture of all the additional tradesmen, along with their equipment and their specific skills, would be needed. But once the right mixture, such as twenty plumbers, fifteen carpenters, thirty electricians, and so on, are acquired, where are they without a set of blueprints to guide them? This is part of the role of the project manager plays. His or her focus is to keep the right people moving in the right direction at the right time, so that the entire project objective is met. By always looking forward and anticipating the outcome, the EPMs goal is to keep the team on course.

But what techniques can an EPM use to attract, motivate, and keep the best and brightest? As the Internet and eBusiness matures, the visions of stock options that convert to fancy cars and riches will soon fade. Many companies use traditional means, such as reward and recognition, training, various perks, more flexible work hours, and a higher level of involvement in business decisions and direction, to name a few. The ability to know what each team member's personal needs are will prove invaluable through out the project.

71

Taking the time to talk and listen to the team is a good habit to have. Many times, e-projects are remotely managed and staffed. Team members can easily get a feeling of being left alone when they are not collocated with the day-to-day project activities and core team. Do not forget about these team members and always be on the look out for signs that a team member is feeling disconnected from the rest of the team. Communication can go a long way for building morale and keeping the team motivated.

Chapter Summary

EBusiness project managers must be able to walk a fine line with stakeholders and team members as they manage the project and its many variables. The project manager must make sure that everyone involved is moving toward a common goal. To help guide the EPM in this function, this chapter covered the areas of assembling an eBusiness project team, setting up a team architecture, identifying and documenting key skill requirements, identifying and communicating with project stakeholders, and keeping the team personnel, thus the project, on track. You are now ready to move forward to Chapter 7, which will discuss defining business requirements.

Chapter 7

Defining eBusiness Requirements

Chapter Preview

During the implementation of an eBusiness solution there are several questions that need to be answered. From what type of browser that will be used to how end users will be authenticated. Defining requirements for an eBusiness solution is not hard and does not require fancy software tools. The keys to success are being able to identify the major components of the project (e.g. technology, business processes); gather information for each component and document the requirements in an easy to read format.

The purpose of this chapter is to provide you with information on how to lead your project team through the process of gathering and documenting the requirements for your organization's eBusiness solution. Some of the concepts that will be covered include the importance of requirements definition; categorization of requirement types and documentation of the requirements.

What is Requirements Gathering and Why is it Important?

"If you don't have requirements, you have nothing to build." Regardless of the type of system, you must define requirements for it. Requirements gathering is the process of gaining insight from multiple sources regarding the expected functionality, look and feel and capabilities of the eBusiness solution. It is an art not a science and requires a high level of precision and detective-like analysis.

The main reason for requirements gathering is to build a mental picture of what the eBusiness solution will be once it is completed. Having this information can help minimize the risk of scope creep and make sure that your project team is spending time on tasks that will lead to the rapid and successful completion of the project within the allotted project time period.

E-Projects move at a fast pace. As a result, the requirements definition process is often performed on an ad-hoc basis or through the use of informal or unstructured approaches. Examples of these approaches include writing requirements for the eBusiness solution on a napkin or a memo pad. Or even worse, recording requirements on a voice mail system. These approaches simply do not work. The main reason for their ineffectiveness is that they lack structure, continuity of thought and are not in a form that can be used to evaluate their impact on other components of the e-project.

Defining requirements correctly the first time is important. First it saves time. Secondly, it saves money. Thirdly, it provides you and your team with a way of auditing the quality of the work that was done. According to surveys conducted by the authors of this book, over 60% of e-projects experience setbacks (e.g. project failure; go over budget or exceed the completion date) because of inaccurate or incomplete requirements definition (See Figure 7.1).

The remaining 40% of e-project setbacks are due to inadequate project staffing, inaccurate selection of e-software and ineffective stakeholder management.

Figure 7.1

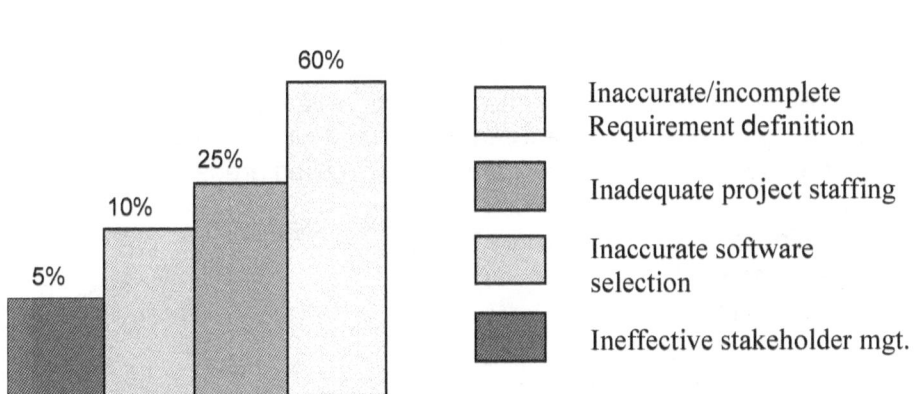

**Major Causes of
e-Project Setbacks**

Based on survey of 75 e-project managers of Fortune 500 companies

What is Different About Gathering Requirements on an e-Project versus a Non e-Project?

Defining requirements on an e-project is very similar to what is done to gather requirements on a non e-project. The similarity is in the process (e.g. sitting down with the project stakeholders and subject matter experts to gather information regarding the eBusiness solution's expected features and functions). One key difference is in the complexity of the requirements that are gathered.

For example on an e-project, when defining content and data requirements, you and your project team will need to know not only what type of content and data that is required for the eBusiness solution, you'll also need to understand the format, size, color, compatibility, structure, source, currency and behavior (e.g. Active X controlled graphics) of the content and data. On non e-projects, gathering content and data requirements is not as complex and detailed because the content and data that is used is often centrally created, stored, maintained and standardized.

As more companies conduct business over the Internet, eBusiness requirements will become complex and dynamic. The challenge for you and your project team is to make sure that the requirements gathering process is focused, concise and is performed with a high level of due diligence and stakeholder participation.

What Requirements Must Be Gathered on an e- Project?

In order to understand how the production version of the eBusiness solution will function your project team will need sit down with the project stakeholders to gather requirements for the following areas:

- Business Process /Business Rules
- Content and Data
- User Interface
- Legal and Regulatory
- Network Infrastructure
- Reporting
- Security and Privacy
- Web Navigation/Page Layout

Business Process and Business Rules Requirements

Business process and business rules serve as the framework for analyzing end user requests and determining how to handle them. It is important to define business process and business rule requirements because it can help you and your project team get a better understanding of what type of transactions will be processed; what type of data is involved; what type of business rules and internal controls are needed.

Business processes and business rules are not the same. A business process is a structured approach used by an organization to manage the flow of information, while a business rule is a decision point within a business process. Figure 7.2 graphically defines the difference between the two terms using a grocery store check out scenario.

Figure 7.2: Business Process Model

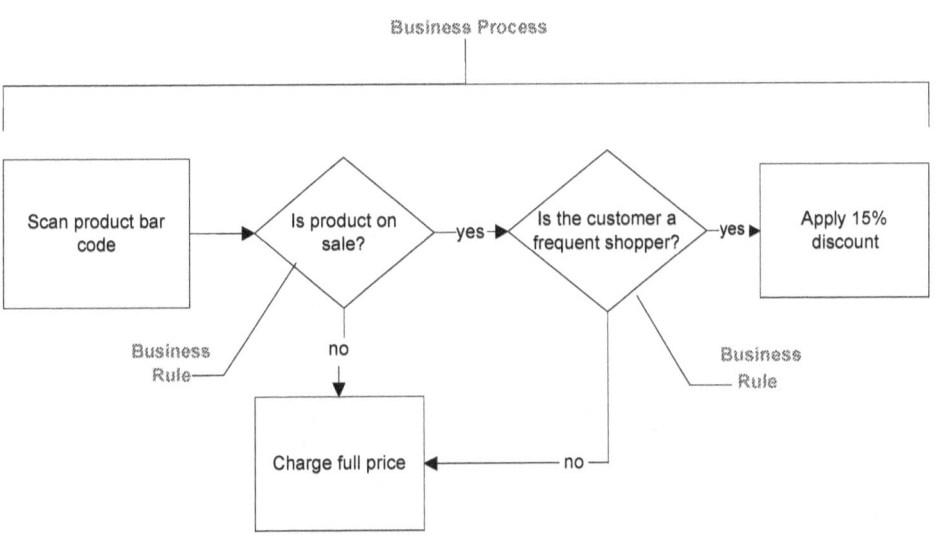

There are two types of business rules: dynamic rules and static rules. A dynamic rule is applied based on the transaction scenario. An example of a dynamic rule is when a customer is given a discount on a product or service once they have spent a certain amount of money or if they patronize a business establishment on a certain day (e.g. "If the order exceeds $1,000.00 discount the order by 15 percent."). A static rule is different. It is applied the same way regardless of the transaction scenario. An example of a static rule is when a customer puts in their personal identification number (PIN) at an ATM machine and the ATM tries to verify the accuracy of the PIN provided. Regardless of the environmental conditions, the application of the business rule does not change.

When gathering business process and business rule requirements for the eBusiness solution, make sure that you and your project team ask questions that shed light on how the eBusiness solution will manage the flow of information from end to end. Examples of some of the questions that should be asked include:

- What processes within the organization will be placed online?
- Who are the owners of each of the processes?

- As information is entered into the system, what types of integrity checks are performed?
- What happens if a transaction fails the integrity check process?
- Where is the information sent once it is received over Internet and passes the integrity check process?

Content and Data Requirements

Content and data are the link between the end user and the eBusiness solution. Gathering content and data requirements is important because it will give you and your project team a better understanding of how the eBusiness solution will be used; what type of user experience will be required and how the information will flow. It will also provide you and your project team with information that will help in getting answers to technical integration questions, such as the type of content that will be used and the technologies needed to design, maintain and publish the content.

Gathering content and data requirements can be complex and challenging. In order to be successful you must engage the project stakeholders and define what is content and what is data.

Content is anything, regardless of form (e.g. HTML, PDF, JPG, text), that has been developed using multiple information sources to communicate a concept or idea to an end user of an eBusiness solution. Examples of content include product literature, online survey form or an e-white paper. Data, on the other hand, is anything, regardless of form, that has no value unless associated with another component of data or data stream. It is used to depict a fact or state of being. Examples of data include phone numbers, account balances and customer names.

Once you have agreed on what is content and what is data, you and your project team should begin asking questions that will paint a picture related to how the content and data will be combined to create a communication pipeline between the end user and the eBusiness solution. Questions that you and your project team should be asking include the following:

- Who will be responsible for creating the content and data?
- What language will the content and data be published in?
- What fonts and formats (e.g. MS Word, PDF, HTML, Text) will be used?
- Who will be responsible for approving content and data prior to publication?

- Are there any personalization issues that need to be addressed?
- How often will the content and data change?
- What process will be used to archive and store the content and data?
- How will content be managed or updated once the solution is in production?

User Interface Requirements

Gathering user interface requirements involves defining what type of experience the end user will have with the eBusiness solution. An example of an end user interface scenario is when an end user enters their user ID and password on the homepage of an eBusiness solution and they are then redirected by the eBusiness solution to a "welcome" page that allows them to perform further actions.

Gathering user interface requirements is important because it will give you and your project team an idea of the type of individuals that will be using the eBusiness solution and some of their characteristics. There are two types of user interface requirements: operational and environmental. Operational user interface requirements are related to how the eBusiness solution reacts to an end user command. An example of an operational user interface is when a separate window opens up when the end user clicks a link on a web page. Another example of an operational user interface is when an eBusiness solution returns a message to an end user once a transaction has been successfully completed. Environmental user requirements, on the other hand, are related to the "soft" features of the eBusiness solution. An example of an environmental user requirement is the background color of the eBusiness solution's web pages or the size and shape of a GUI enabled button on the toolbar section of an eBusiness solution's homepage.

Some of the questions that need to asked during the user interface requirements gathering process include:

Operational User Interface Requirements

- What type of messages should be returned to the end user once a transaction is submitted for processing?
- How should the message be presented to the end user (separate window, redirection to a home page, automated email)?

- What language should the message be presented in to the end user (e.g. French, English, Spanish)?
- How long should the end user be given to respond to a message that has been generated by the eBusiness solution?
- Where and how will the end user enter their response?
- What size should data input fields be?

Environmental User Interface Requirements

- Who will be accessing the eBusiness solution and what is their educational and social background?
- What color and size should the screen be?
- What shape and color should the hyperlinks and buttons be?
- How should the buttons or links behave when the cursor is placed over them?
- Where should the navigation menu be located (across the top or down the left side)?

Legal and Regulatory Requirements

Gathering legal and regulatory requirements involves the process of capturing information from the project stakeholders that addresses functionality and features of the eBusiness application that will need to adhere to regulatory and legal guidelines.

It is important to gather requirements related to legal and regulatory issues because it will ensure that your client organization is in compliance with any applicable laws or standards. It will also minimize the possibility of being penalized by an authoritative body or sued by a customer.

Legal requirements are not the same as regulatory requirements. A legal requirement is a statute that has been put in place by a law setting organization and applies to an entity regardless of what industry it competes in, while a regulatory requirement is a statute that has been put in place by a standards setting body and only applies to organizations that fall under the authority of the standards setting body. An example of a legal requirement is the Graham-Leach-Bliley Bill of 1999 that addresses protection of an individual's privacy. An example of a regulatory requirement is, the Internet Banking Regulations that were developed by the Office of the Comptroller of Currency to provide guidance to the financial services industry (e.g. banks, credit unions), on how to protect customer information within an e-banking environment.

Some of the questions that you and your project team should ask when gathering requirements related to legal and regulatory issues should include the following:

- Are there any international laws or regulations that the organization must adhere to?
- Are there any local laws and regulations that the organization must adhere to?
- Who is responsible for making sure that all applicable guidelines are adhered to?
- What techniques are used to monitor compliance with the applicable laws and regulations?

Network Infrastructure Requirements

The network infrastructure of an eBusiness solution serves as the back bone of an eBusiness solution. It is made up of servers, routers, sub-networks and software applications. Understanding network infrastructure requirements is important because it will help you and your project team decide how to design and configure each layer of the network infrastructure. It also helps in determining what system integration strategy to use.

There are two approaches used to integrate the different components of an eBusiness solution network infrastructure: point-to-point and many-to-many. The point-to-point approach involves making connections from a single system to a single system over and over again by using different point-to-point middleware. The many-to-many approach involves the middleware software connecting to different computing platforms. While currently over-hyped and under-tested, products like message brokers, allow enterprises to connect many diverse systems, typically without changing the source and target systems.

Other popular integration approaches include database-to-database, multiple databases, and brokered systems. The database-to-database approach integrates applications by connecting the databases. It can be done at the database level when information sharing between applications is light. The multiple database approach integrates applications by accessing many databases as one logical database or through a single API. The brokered systems approach integrates applications by brokering information (messages) moving between them. It can be used when many different applications must be connected and information sharing must occur at the database and process level. It works the best when major changes to the source or target systems are not desired.

Some of the questions that you and your team should ask project stakeholders regarding the network infrastructure requirements include:

- What are the performance requirements for the eBusiness solution (e.g. response time, application availability, and thru-put)?
- What will the integration requirements be?
- How much traffic will the eBusiness solution have to support (peak load and average load)?
- Where will data that is processed by the eBusiness solution be stored/archived?
- What type of connections will the eBusiness application support (e.g. DSL, Cable Modem, Dial-up, T1)?
- How often will the eBusiness solution need to be backed up?

Internet Security and Privacy Requirements

Gathering information regarding Internet security and privacy requirements involves the process of determining how information within the eBusiness solution will be protected. Having a good understanding of the security and privacy requirements for the eBusiness solution is important because it will help you and your project team get a handle on potential security and privacy threats that may exist. It will also help in identifying what other components of the project may be affected (e.g. technology architecture).

The threat of security and privacy violations is real. Companies, over the last several years, have lost billions of dollars in sales and customer goodwill due to security and privacy breaches. The key to successfully gathering security and privacy requirements is to first define what security is and what privacy is. This is the best approach because the two terms are often used synonymously but they are different. Security involves the use of Internet technology to manage access to information assets. Privacy is different. It involves the use of Internet technology to manage how information assets are used.

Some of the questions that you and your project team will need to ask the project stakeholders include the following:

Security

- How will the identity of an end user be verified?
- What type of access will end users have and how will it be managed?
- How will end user transactions be monitored?
- How will security breaches be identified and resolved?

Privacy

- What information will be entered into the eBusiness solution?
- Who will use the information?
- When will the information be used?
- How will the information be used?

Reporting Requirements

One of the key eBusiness solution requirements is reports. A successful eBusiness solution is one that can deliver the reports desired by an organization. An eBusiness solution needs to generate various kinds of business reports that can be used by end users and project stakeholders to make business decisions. Some examples of the type of reports that you and your project team may be required to develop include web statistics reports (e.g. number of hits, number of sessions), online sales reports and online survey reports.

It is important to gather reporting requirements because it will help you and your project team better understand how each business process works from an end to end perspective.

There are two types of reports: structured and unstructured. A structured report contains data that is displayed in a tabular or a diagram format and is based on a pre-defined query. An unstructured report contains data that is displayed in multiple formats and often has to be manually generated using custom developed queries.

In order to get a clear understanding of the reporting requirements, some of the questions that you and your project team will need to ask include:

- Who is the user of the report (customers, managers, staff)?
- How often will the report need to be generated?
- What method will the report need to be delivered in (email, HTML, XML)?
- What information will be on the report?
- What will be the layout of the report?

Web Navigation/Page Layout Requirements

Web navigation and web page layout requirements define how an end user will get from point A to point B while using the eBusiness solution and how the information will be displayed to them. Understanding web navigation and web page layout requirements are invaluable because it enables you and your project team to better understand the end user's experience and how to go about designing the most efficient web navigation architecture and web page layout.

There are three types of web navigation design approaches: dashboard, integrated text and consolidated. The dashboard approach manages how an end user navigates through the eBusiness solution through the use of a static menu or tool bar that shows up on every web page. The integrated text approach allows an end user to navigate through the eBusiness solution by clicking on key words that link to other web pages within the eBusiness solution. The consolidated approach is a combination of the dashboard and integrated text approaches. It allows an end user to navigate through out the eBusiness solution by either clicking on linked text or on a menu button that is placed on a tool bar.

There are two types of web page layouts: horizontal and vertical. The horizontal layout is designed to display content from the left side to the right side of the web page. The vertical layout is designed to display content from top to bottom. Both layouts can be effective. Determining which one to use is a function of how much content and data that will be posted on each page.

Some of the questions that you and your project team will need to answer include:

- How many clicks should an end user perform in order to get to the information they desire?
- What information will be on each web page?
- Where will the information be placed on the web page?
- How many entry point will the eBusiness solution have?

How Should Requirements Be Gathered and Documented?

Gathering and documenting requirements is essential. In order to make sure that all the requirements are captured and documented, you and your project team will need to develop an information capture and documentation process. The information capture process should involve someone from your project team (e.g. business analyst) sitting down with a subject matter expert (SME) from the client-side for a question and answer session (Q&A). The time spent with the SME will require direct and focused communication. The questions asked should be open ended to allow the SME to be able to provide a comprehensive response (e.g. "Who will be the users of the eBusiness solution once completed?").

The responses provided by the SME must be documented in an easy to read format. One of the best ways to do this is to use a checklist that provides a space where the interviewer can add their comments to the responses provided by the SME. Figure 7.3, found in the reference section in the back of this book, provides an example of an end user security checklist that supports this documentation approach.

Chapter Summary

The focus of this chapter was to help you understand what goes into defining requirements for an eBusiness solution. The task is not difficult. It just requires that you and your project team have a structured process that is keenly focused on the different components of an eBusiness solution. In chapter 8 you will learn how to take the information that was gathered during the requirements definition phase and put it into a format that graphically depicts what the eBusiness solution should resemble once it is completed.

Chapter 8

Designing the eBusiness Solution Architecture

Chapter Preview

Designing the eBusiness solution architecture is a structured process that involves taking the information that was gathered during the requirements definition phase and putting it in a format that depicts how the requirements will be met. The scope of the design should include all pertinent areas of the eBusiness landscape (e.g. business process, content and data management, security). The eBusiness design process must include all project stakeholders (e.g. project team, business partners, and client end users.) The result of the time that you and your project team put into designing the eBusiness system is a document that explains how the eBusiness solution will work-from top to bottom.

The purpose of this chapter is to take you through the architecture design process. After reading it you will be able to evaluate the quality of work that your project team performs on your eBusiness engagement during the design phase of the project. You will also be able to answer any questions they may have regarding the architecture design process.

What is Involved in Designing an eBusiness Solution Architecture?

Designing an eBusiness solution architecture involves reviewing your deliverables and notes from the requirements phase and determining how to architect the system so that it performs the way the project stakeholders would like it to.

An example of some of the design related activities that you and your project team will perform include developing process flow diagrams that depict how customer orders are processed. This effort will include architecting how the price is calculated; how payment is collected; and how shipment of the product is handled, just to name a few.

The design process involves working with individuals at all levels of the project stakeholder network. For example, in order to properly design the online order entry process, you and your team will need to work with stakeholders in the inventory management, credit and finance, billing and collection, customer service, technology support and legal and regulatory areas. This is just an example. Depending on the complexity and scope of your e-project, the number of stakeholders may vary.

What is different about designing an architecture for an eBusiness solution versus designing an architecture for a non eBusiness solution?

The process of designing an architecture for an eBusiness solution is very similar to designing the architecture for a non eBusiness solution. The difference is in the number of different access points that an end user can use to utilize an eBusiness solution; the diversity of an eBusiness solution end user population and the complexity of the technology components that make up an eBusiness solution's backend processing capability.

End user access points

End user access points refers to the way an end user accesses an information system. EBusiness solutions have access-touch points that are not commonly found in non eBusiness solutions. For example, an eBusiness solution can allow end users to access information assets using multiple communication protocols (e.g. FTP, HTTP, Telnet, WAP). It can also be designed to allow them to access information using multiple devices such as a wireless phone, personal digital assistant (PDA), pager or laptop computer. This is different from a non eBusiness solution where in many cases the communication protocol and access channel is standardized or limited.

Diversity of the end user population and personalization

End user diversity and personalization refers to the variety of end users that can require access to an eBusiness solution. Designing end user personalization architecture can be a challenge because it must be flexible enough to recognize when an end user' preferences have changed, but rigid enough to maintain the integrity of the transaction process.. An example of end user personalization can be found on the websites of some of the large e-tailers (e.g. walmart.com). In some cases the e-tailer's eBusiness solution must be able to not only process customer orders, but it must also be able to customize how information is presented to the end user based on customer preferences (e.g. product interests). Designing the end personalization process is more straightforward for a non eBusiness solution because the personalization capabilities of the system are often standardized and structured.

Technology Architecture Design Complexity

Technology architecture design refers to the process of architecting the components of the eBusiness solution that will automate the information handling process (e.g. web servers, routers, application servers). The difference in designing the technology architecture for an eBusiness solution versus a non eBusiness solution is that eBusiness solutions can be built using a wide array of software and hardware technologies that often do not integrate well. As a result, the design process involves developing mini applications that can allow the solution to perform properly. For example, when digital certificates are used for end user authentication, the project team does not only have to design the certificate administration process (e.g. enrollment, revocation, expiration), they also have to design an architecture that ties the authentication server to the certificate key server (the server that holds the private and public keys of the end user) of the certificate authority (the organization that is responsible for distributing digital certificates). For non eBusiness solutions architecture design is not as difficult because standards have been put in place to ensure that data sharing is enabled. Also, non eBusiness solutions often do not require information gathering over dispersed-heterogeneous systems that are not under the control of the user organization.

Who is responsible for designing the eBusiness Solution Architecture and what deliverables are they responsible for?

The responsibility of designing the eBusiness solution rests with you, your project team and the project stakeholders. This is a joint effort due to the fact that the project stakeholders often know more of the detail behind how some of the processes work and how the technology architecture functions.

The deliverable from the project team should be a document (i.e. eBusiness Solution Design Document) that defines how the requirements of the project stakeholders will be met. It should provide detailed information related to how the following components will be designed:

- Business Process
- Content and Data Management
- Legal and Regulatory
- Network Infrastructure
- Reporting
- Security and Privacy

Business Process

Designing the business process architecture for an eBusiness solution involves defining what steps and decision points need to be performed to ensure that a transaction or end user request is properly handled. The business process architecture should be the first area that is completed during the eBusiness solution design process because it determines how processes will function (See figure 8.1).

When reviewing the work that your project team has done on defining the business process architecture, make sure that the process considers all steps from an end-to-end perspective. That is, it should define how, when and where a transaction is created, analyzed/approved and integrated with other parts of the eBusiness system. It should also define how, when and where the transaction is archived. Figure 8.1 is an example of how the business process design is linked to the other project components.

Figure 8.1

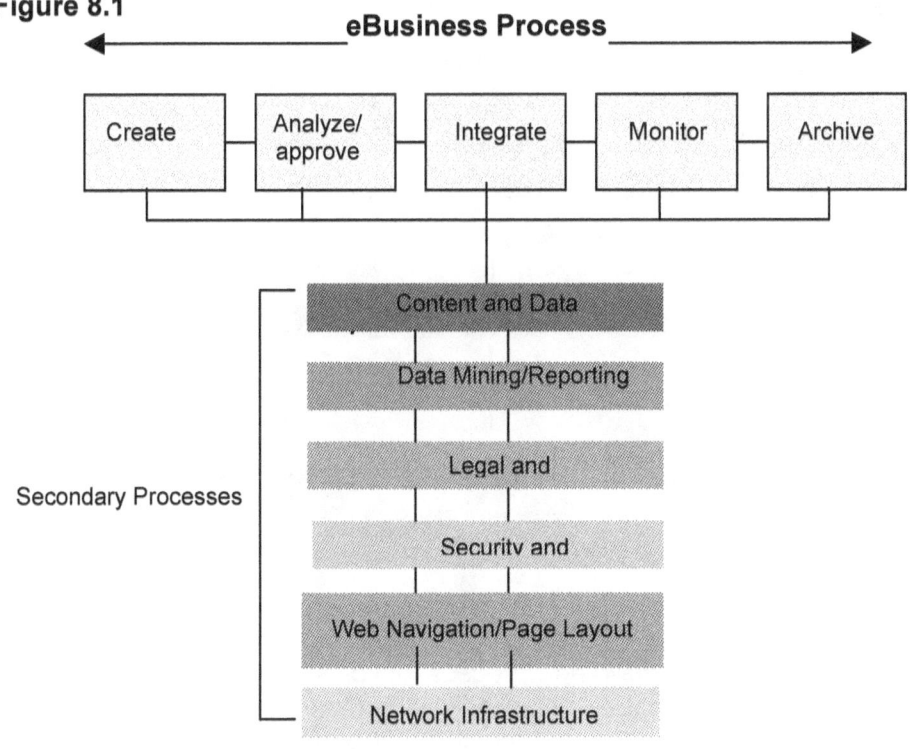

Content and Data Management

Designing the content and data management process architecture involves putting together an architecture that addresses how information is created, approved, published and archived (See figure 8.2). In the eBusiness world, content and data can take many forms, therefore, it is important to make sure that steps involved in content and data creation clearly define the different types of content and data that will be used. When reviewing the approval process make sure that your project team clearly defines who is responsible for approving the content and data before publication-placing the content and data on to the Internet. The design should also define how, when and where content is published (e.g. top of the web page). This is important because many organizations have users that are in different time zones and require that the publication process be managed more effectively. Finally, the content and design process should define how the archival process will work (e.g. what is archived, when it is archived, what format/structure it is archived in).

Figure 8.2

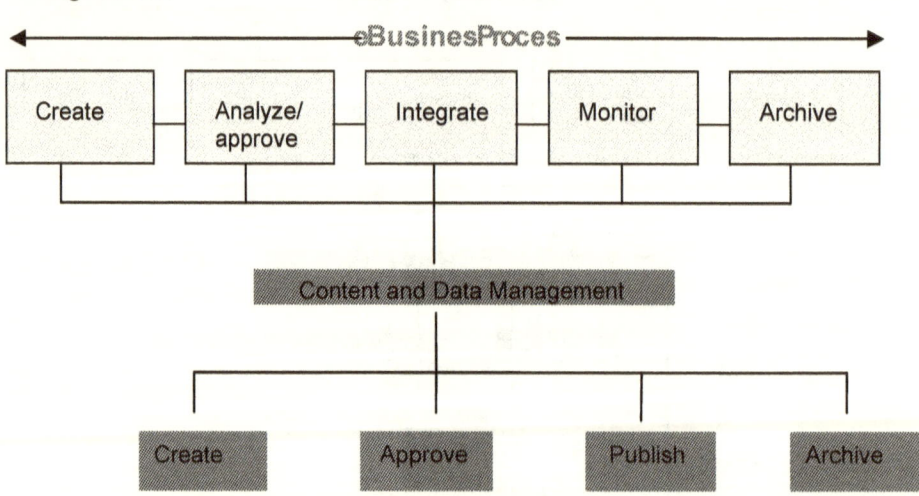

Reporting

Designing the reporting process involves putting together an architecture design that explains how information is extracted from the eBusiness solution. It is important to understand the design of the reporting architecture because it will determine what information can be extracted; what format it can be displayed in; and who will be able to access the information.

When reviewing the work that your project team has done to design the reporting process, make sure that they define the steps that need to be performed in order to select, generate, view and archive reports. The report selection process should define what type of reports are available; what each report is used for; what format or structure the report is in; who are the users of the report; how an end user selects a report and what are the sources of information for the report. The report generation process should define how an end user goes about generating one of the available reports, while the process design related to viewing the report results defines what options are available to the end user to view the report (e.g. online vs hardcopy). The report archival process should define how, when and where the reports are archived. Figure 8.3 shows how the reporting process architecture are linked.

Figure 8.3

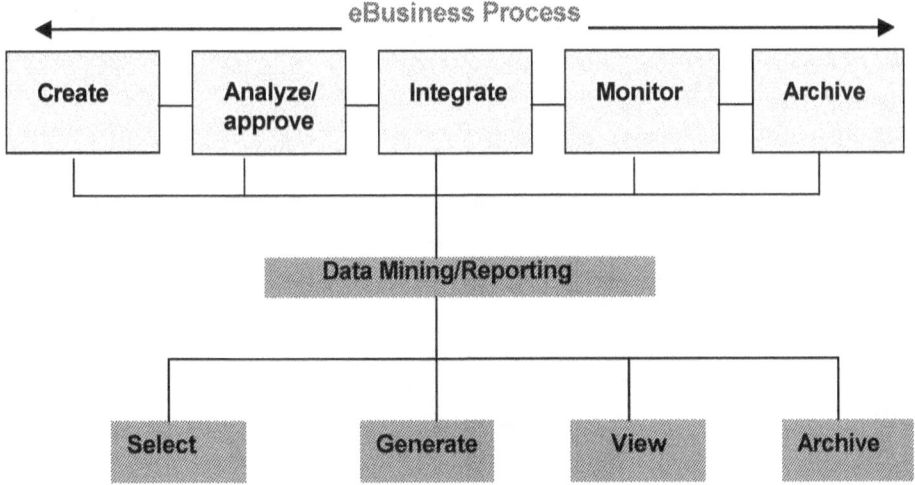

Legal and Regulatory

Designing the legal and regulatory evaluation architecture involves examining the requirements and standards that the user organization has to adhere in the offline environment and then design a process that shows how it will be done in an online environment. For example, an organization that ships hazardous materials must adhere to certain standards that have been put in place by the Department of Transportation (DOT). When the same organization develops an eBusiness solution where customers can place orders over the Internet, it must develop a mechanism for determining how to make sure that the DOT requirements are not overlooked.

The key to success in designing the legal and regulatory process architecture is to automate the evaluation process similar to what has been done with business rules and end user authentication. When reviewing the legal and regulatory process architecture make sure that your project team defines how information related to the legal and regulatory framework is gathered; how it is evaluated and how the response is provided (automated or manual). Figure 8.4 is an example of the components that are involved in designing legal and regulatory process architecture for an eBusiness solution.

Figure 8.4

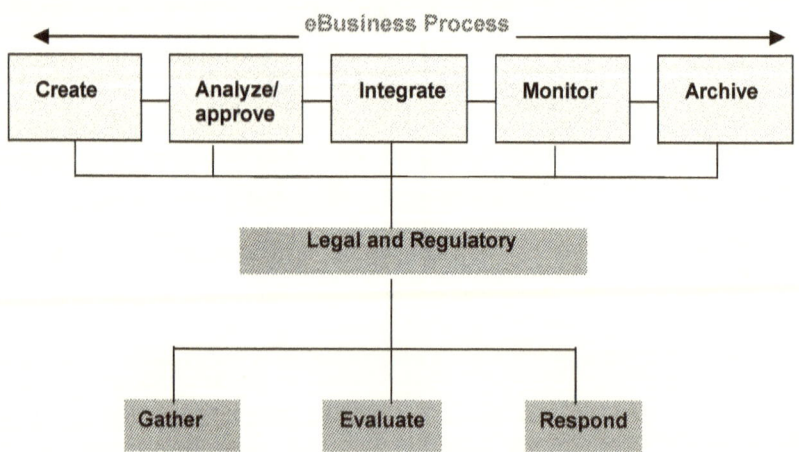

Security and Privacy

The design of the security and privacy process architecture is very important. The reason why it is important is primarily due to the fact that it is your organization's main defense for preventing unauthorized users from gaining access to the eBusiness solution.
When reviewing the security and privacy architecture design that your project team has put together make sure that it clearly defines how and when end users are authenticated; what type of authority is assigned to each user; how end user activity will be monitored. Also make sure that the project team has defined how access rights will be managed on an on-going basis.

Figure 8.5 provides a graphic example of what elements should be a part of the security and privacy process architecture.

Figure 8.5

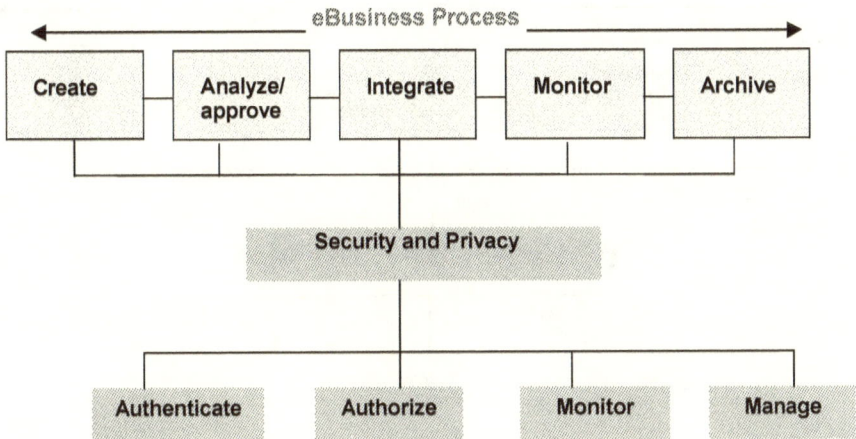

Web Navigation and Web Page Layout

The web navigation and web page layout design architecture involves having your project team define how an end user will navigate through the eBusiness solution and where the content and data will be placed on the web pages of the eBusiness solution. It is important to understand the navigation and page layout design architecture because it may expose areas of the architecture design that have not been properly integrated.

When reviewing the architecture design for the web navigation and page layout area make sure that your project team has clearly defined the names, entry points, exit points and browse features for each web page. Entry points represent points within the navigation architecture that can be used to go to a specific web page, while exit points represents areas within the architecture of the page that will allow an end user to leave a specific web page. Browse features represent points on a web page that allows an end user to navigate to different points within the same web page. Figure 8.6 provides a graphic depiction of how the different components of the web navigation/web page layout architecture are linked.

Figure 8.6

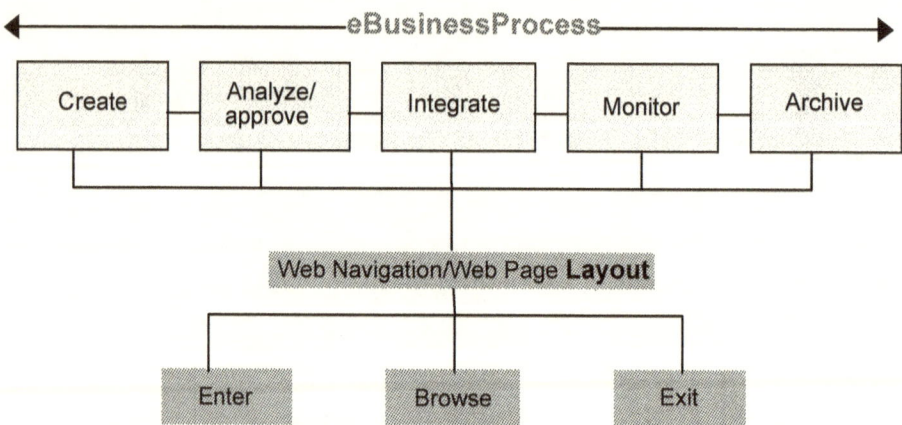

Network Infrastructure

Designing the network architecture involves putting together and integrating the technology components (software and hardware) in a way that the eBusiness solution performs as expected. There are two types of network infrastructures: static and dynamic. A static infrastructure is built to support a limited number of features and functions. It is also designed to limit what interface with a limited number of other systems and devices. An eBusiness solution that is developed in HTML; does not integrate with any other systems and only can be accessed using the hypertext transmission protocol (HTTP) is an example of a static network infrastructure.

A dynamic network infrastructure is built to support several features and functions. It is also designed to interface with multiple devices using multiple communication protocols. An eBusiness solution that is developed in XML; provides end user access using either HTTP, file transfer protocol (FTP), or wireless access protocol (WAP) and can accept data uploads from an assortment of wireless devices is a good example of a dynamic network infrastructure at work.

When reviewing the architecture design that your project team has put together, make sure that the proposed architecture is designed to use a limited number of system-to-system integration connections; avoids server traffic overloads and minimizes client-side customization requirements. Also, make sure that the project team has also defined the architecture from a functional and technical perspective.

The functional perspective of the network architecture should define what components are required to enable the eBusiness solution to function properly. It should also explain what each component is used for. An example of a functional design is when developer documents how the database, web server, application server and the router will perform when an end user is authenticated.

The technical design of the network infrastructure defines how the components will be connected or linked. It also defines the technical specifications that are required for each component. To get a clear understanding of how the network infrastructure will look once it is built, make sure that your project team provide a graphical depiction of the network infrastructure (See Figure 8.7).

Figure 8.8

How should the architecture design process be documented?

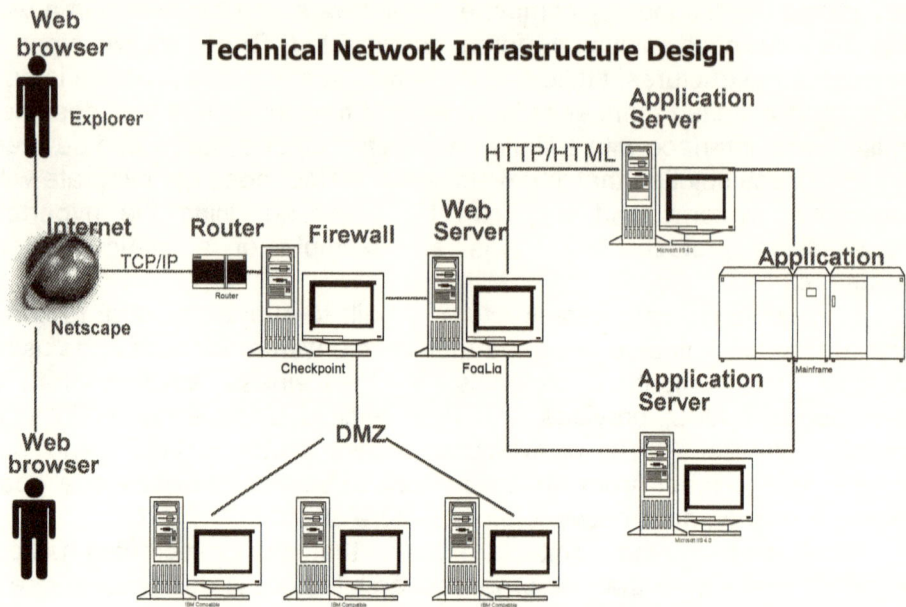

There are various ways of documenting the design architecture. One way is to use a flow charting software. Another approach is use a text-based approach along with flow charts. The key to documenting the architecture for the eBusiness solution is not in the tool that is used, but the accuracy, completeness, validity and relevancy of the information that is compiled. There are software tools that are available from Microsoft (e.g. Visio) and Rational Software (e.g. Rational Suite DevelopmentStudio). When evaluating a documentation or modeling tool make sure that it can support your functionality requirements and allow for sharing information with all project stakeholders.

Chapter Summary

In this chapter we walked you through the process of designing the eBusiness solution architecture. The effort involved in the design process is very important. Without going through this exercise, your project teams will be lost when it is time to "build" the architecture. In the next chapter- chapter 9, we will explain how to go about building the eBusiness solution architecture and managing the customization process.

Chapter 9

Building and implementing an eBusiness System

Chapter Preview

The building and implementing phase of an eBusiness project involves the installation, customization, unit testing, and integration testing of the system (application) modules, screens, and reports. It will also address the preparation and establishment of the technical environment for development, testing, and training of user representatives. The EPM has overall responsibility to ensure that the deliverables associated with the detailed implementation design are produced.

This chapter will discuss the tasks involved with this phase of an eBusiness system. It presents the general order of installation and configuration tasks, how the tasks relate to each other, and where to find the information needed to complete them. Some of the topics discussed in this chapter, at first glance, may look out of context, but in reality that is not the case. Much emphasis is given to explaining the architecture in simple terms. There is a definite need for the EPM to have a level of understanding about these various phases to be able to validate and direct the project team.

The tasks involved will include: planning configuration, setting up databases, installing and configuring the eBusiness software components, configuring the existing systems to work with eBusiness systems, and testing. A detailed explanation of testing standards is provided at the end of the chapter for the QA (Quality Assurance) part of an eBusiness project.

Planning The Configuration

Planning the configuration involves the arrangement and documentation of hardware and software configurations and verifying that these configurations meet the business needs identified by the organization. There are many factors that affect the selection of the right technology for eBusiness: architecture, configuration, and hosting issues are key to a successful implementation. As an EPM you will need to have some understanding of each these phases. The depth of understanding will be up to the individual as this is determined by your previous

98

experience and the level of talented individuals you have on your team. Lets look at each of these in more detail in the following paragraphs.

Architecture

Understanding the eBusiness architecture is the foundation to understanding the remaining components of an e-project. Architecture is the components that makes up a system. It also describes the relationship between those components. The architecture of a system can be simple or complex, depending on the nature of the system or business. The architecture for a business-to-consumer system might be very different from a business-to-business system. This section will discuss the main components of eBusiness architecture and characteristics of the architecture.

Main Components of eBusiness Architecture

The main components of eBusiness architecture include: Web browser, Web server, e-commerce engine, database server, order fulfillment, payment gateway, data refresh, and external systems. See Figure 9.1.

Figure 9.1
Main Components of eBusiness Architecture

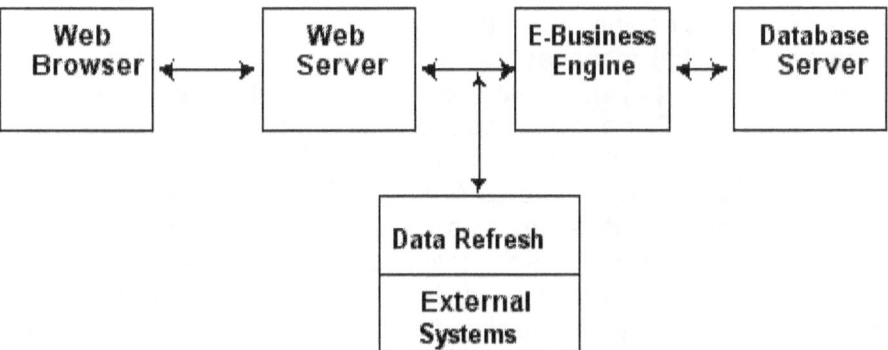

Web Browser

The Web browser is the user's most important software interface. It's the software that enables the user to see HTML pages that make up the eBusiness solution. Therefore, it's important for the solution to only use features that the user's Web browser is capable of displaying or to provide ways around the feature.

For example, some browsers are not capable of displaying frames, therefore you may need to provide the option of viewing the Web page without frames. Otherwise, the user's will not be able to see the content in the frames. For security purposes, it is important for the user browser to support Secure Sockets Layer (SSL) sessions. The take away for the EPM is to make sure compatibility is documented and communicated. There is more about security in a following chapter.

Web Server

The Web server acts as the go-between for the eBusiness engine and the user's Web browser. When the user clicks a link on the Web page, for example, the command is sent to the Web server, which then forwards it to the eBusiness engine. The engine acts on the command and then sends the information back to the Web server. The Web server will also store the catalog (HTML) pages that the user's Web browser will display.

EBusiness Engine

The eBusiness engine or server is the heart of the eBusiness system. It's the software that manages the Web server, database server, order fulfillment system, payment gateway, data refresh system, and the external systems. The engine receives input from the Web server. The engine then retrieves information from the database and provides input back to the Web server. The EPM should have a sense of how the eBusiness engine manages various servers and external systems so that he or she can identify the right team lead or technical lead should problems arise. Your responsibility is not to be able to trouble shoot problems, but to identify and resolve problems by focusing the right team members on the issue. All three of us (the authors) firmly believe that for every eBusiness project there should be a project manager who is not the technical manager. This allows the technical team to focus on what they do best and allows the EPM to manage the project.

In this next section we will be discussing eBusiness architecture, payment gateways, scalability, reliability and other relayed topics to architecture. The EPM's role in this part of the project would be to ensure the business requirements that the stakeholders have identified are truly being addressed. Sometimes and without intent, these discussions can wonder. The EPM must keep the team focused and moving toward the end result. Not to be contradictive but, it is also the EPM's responsibility to stop the project if after reasonable discovery it is determined that the end result cannot be achieved or the resources need to complete the project have been either re-deployed or are not available. Do not be afraid to stop a project if you know it will fail.

Database Server

The database server is the most important feature of a eBusiness system. This is where the eBusiness solution stores actual data. The success of eBusiness depends on the powerful database system that allow the users to have fast, secure access to all the data. The idea is to create one database system that handles every function of the eBuisness solution. All popular databases are offering the solutions designed for typical eBusiness features. The EPM must take a closer look at all the features of the database before selection.

Data Refresh

To attract new buyers and keep current ones, the merchant's online catalog needs to constantly change. For example the catalog of an eBusiness website needs to show price changes, discontinued products that have been removed, and new products that have been added. This is all great except what does it have to do with me the project manager? Tactically it has very little, but strategically it has a big impact. Site maintenance and content refresh are two of the most overlooked areas in any eBusiness project. Your responsibility to the client is make sure they understand the impact and implications of post implementation support for the site. Especially if you are developing, hosting and supporting the entire project. Either of these can be time consuming and large efforts considering the frequency of refresh and change that is required by the client.

External Systems

External systems are add-on utilities that did not come with the original e-commerce system. The external system might provide a function, such as converting currency at the point of purchase, that the original system was not able to perform. The EPM take here is to make sure that these systems are documented, tested, supported and are systems that everyone agrees will perform and scale as expected.

Characteristics of eBusiness architecture

The following are the important characteristics of the architecture of an eBusiness system:

- Scalability
- Extensibility
- Flexibility
- Manageability
- Reliability
- Recoverability
- Availability

Scalability

EBusiness architecture should have the ability to grow and provide acceptable levels of performance for multiple users at high transaction volumes. Scalability can be provided at the hardware, software, and application level. A general rule of thumb when deciding how much scalability is needed is to "plan for success". It is much easy to scale down than up.

Flexibility

A good architecture model should be able to accommodate and incorporate new technologies as they become available. Flexibility is a critical characteristic of an e-commerce system architecture. The architectural design should also ensure that the features and services provided are accessible from GUI clients, Web browsers, voice response units, and other clients. Your goal as an EPM is to confirm that the system will be able incorporate these new technologies without monumental efforts or cost.

Reliability

Your role as the project manager would be to ensure that the architecture provides dependable support for its users. Additionally, as the EPM, your role should be to define what reliability means based on the input of the users regardless of weather or not they are internal or external. It should support Atomicity, Consistency, Integrity, and Durability (ACID) of transactions across multiple resource managers (legacy applications, Enterprise Resource Planning (ERP) applications and departmental databases).

Availability

The architecture supports operational capabilities that are consistent with the Service Level Agreements, (a contract between provider and a customer that specifies, usually in measurable terms, what services the provider will furnish) established for this application. For example, if the supporting infrastructure becomes unavailable, it should be able to recover and restart the applications within acceptable predefined time frames. Your role as an EPM would be to make sure that the Service Level Agreement was in place and all parties have agreed to the terms of the agreement.

Extensibility

The architecture should have the ability to be easily enlarged to support new business functions. It should also support easier upgrades of the underlying hardware and software. Once again, as the EPM you are the point of contact to coordinate, facilitate and document what these new functions will be, or at the very least find out who would be in a position to define new functionality.

Manageability

The architecture is supported by a set of commercially available tools that will simplify the systems management functions for the distributed components of the architecture. This includes starting and stopping services, automatic restarts, monitoring and error reporting, software distribution, and performance tuning.

Recoverability

The architecture should be able to provide error detection and recovery functions for all of its services. It should also assure that transactional and database integrity is preserved through the restart and recovery process. All technically sound direction for a robust system. Leave these types of decisions to your lead architect, that is what they are getting paid for. Your goal is to get them to put the commitment in writing and make sure it meets the requirements.

Configuration

Basically, there are three different configuration approaches: single system, multi home single system, and multi machine. We are providing an overview of each of these for the benefit of those EPM"s who may not have a technical background.

Single-System Configuration Approach

A single package in a single system is the simplest set up for an e-commerce system. In this configuration, one e-commerce package is installed on one machine with one server. The three main components of a single-package, single-machine configuration are the database, the e-commerce server, and the Web server. See Figure 9.2

Figure 9.2

Single-System Configuration Approach

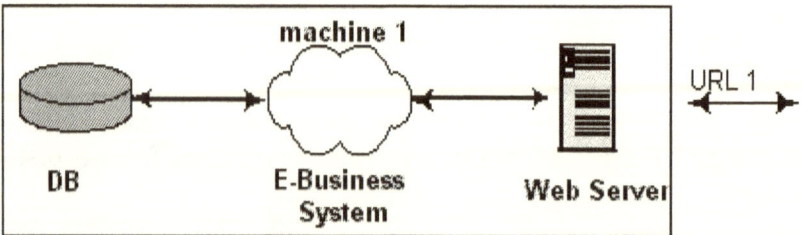

104

Multi home, Single-Machine Configuration Approach

A multi home, single machine configuration occurs when one e-commerce installation is used for several shops or malls. In a multi home, single-machine configuration, two or more e-commerce servers are each connected to a database and Web server. Testing these applications for load, stress and integrity are a must prior to going live. See Figure 9.3.

Figure 9.3

Multi home, Single-Machine Configuration Approach

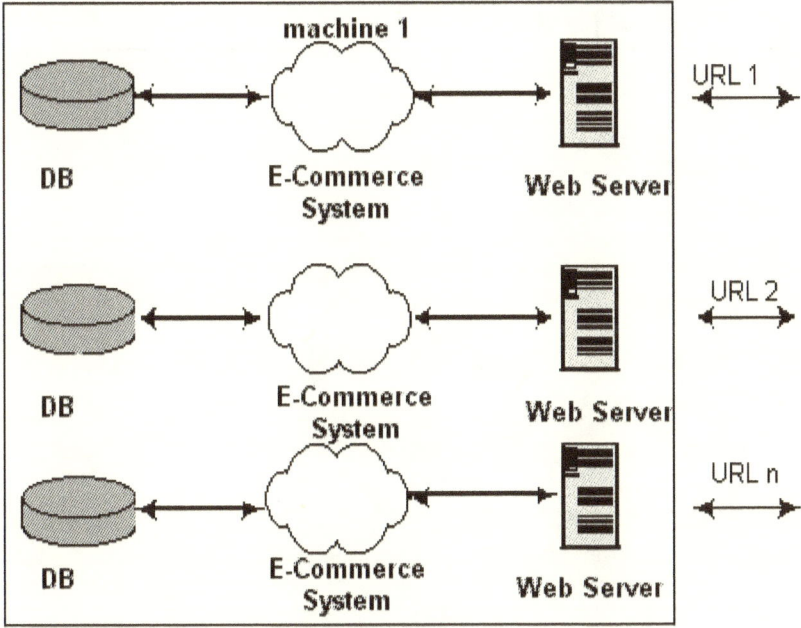

Multi machine Configuration Approach

In a multi machine configuration, the database is in one machine, while the e-commerce system and Web server are in another. One advantage of using multiple machines is that the Web server and the database load can be separated over two different machines.

Figure 9.4 is an example of a two-machine configuration where a database client is attached to a machine that contains the e-commerce server and the Web server.

Figure 9.4

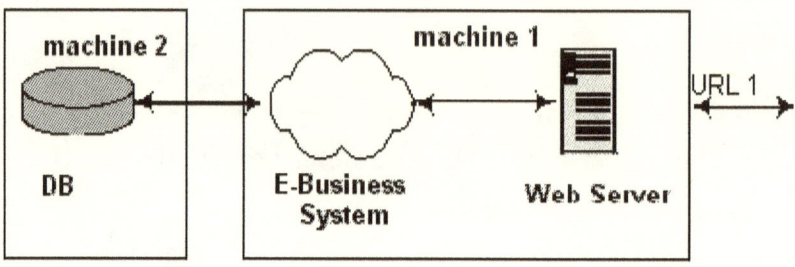

Figure 9.5 is another example multi machine configuration. In this example, two or more machines, each with an e-commerce system and a Web server, are connected to another machine controlling a database at the back end and load balancing system at the front end. A load balancing system, such as a Network Dispatcher, allows the customer to see only one address and one server. This configuration provides an effective way to operate multilingual sites.

Figure 9.5

Multi machine Configuration Approach #2
11

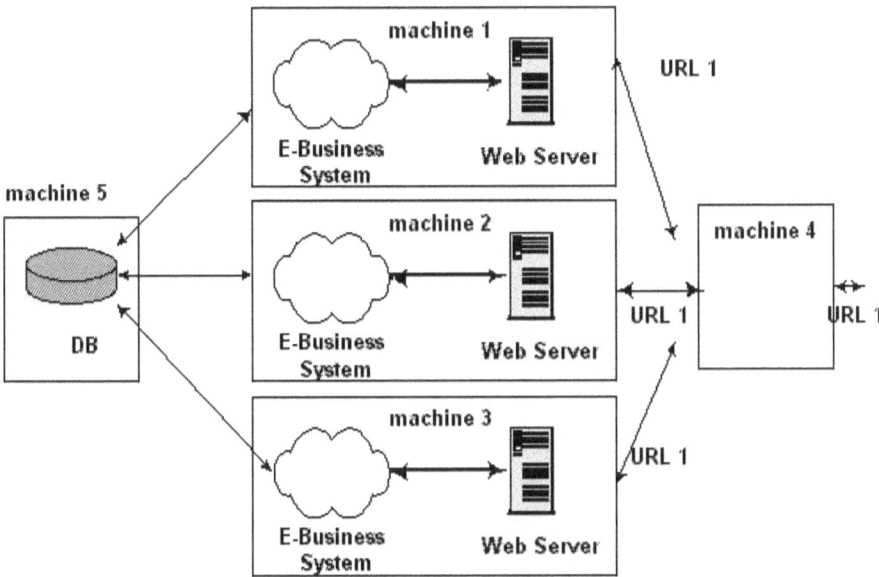

Hosting Issues

The merchant might already have a host system to use to set up the e-commerce system. If that is the case, the system must run on the merchant's computer system. This issue alone will determine which e-commerce package the merchant can use.

If the merchant's site is going to be hosted by a third party, it is important for the merchant to address security and connectivity issues early. For example, the merchant will need to verify whether or not security policies would conflict with the party hosting the site.

Enter the EPM, this is where the work begins when trying to coordinate between two systems that may or may not be compatible or may not be in your control. The EPM plays an important role as mediator and is looked upon to ensure that there is a communication taking place between both parties and progress is being made. Agreements along with acceptable behavior expectations and problem resolution between either company should be lead by the EPM with support from the technical team, legal and the contracts department.

Setting Up the Databases

In setting up the databases, you should assure that the technical team completes the proper documentation for database planning and that the growth plan is in place and documentation is maintained for the task. Depending on the size of the company and eBusiness project you are working on, a database architect, developer or lead should be engaged and assigned to the project. Your role as an EPM would be to identify resources, assist with team selection and to work the business side of the process.

Seldom, if ever will a database architect work through the entire project. Once the aspects of the architecture have been defined, or at some point after that, a database programmer, analyst or other database specialist will usually complete the project from that point. Coordination, documentation and oversight are the responsibility of the EPM. Making sure that during any transition that neither time or project continuity is lost is also the responsibility of the EPM.

Installing and configuring the eBusiness Software Components

Before installing and configuring an eBusiness system, the EPM should assure that the hardware and software configurations meet the requirements, selected the organizational architecture, set up the databases, and completed the Configuration Worksheets. Work in this phase for the EPM is to document and distribute this information so that all of those who need to see, read, understand, test and validate this information are engaged. Merely sending a brief e-mail out to the team with attachments hoping they find time to read it would be a costly mistake for the EPM. The responsibility here is for the EPM to identify the stakeholders who will be effected and to engage them. It would not be the responsibility of the EPM to validate the configuration for themselves, unless the project was very small and the EPM was confident in their own skills and knowledge to endorse moving forward.

Configuring the Existing Systems with eBusiness Systems

The eBusiness system always coexists with the other systems in the organization. This demands for true integration of different technologies.

The configuration issues in many cases can be significantly high and many times complex. Proper planning and coordination helps to avoid transition disasters.

One of the many configuration integration challenges that the EPM will face is how to integrate with legacy systems and present information. The successful EPM will have a good understanding of these systems and what roles they play in the business. The EPM role here is to manage this integration and to insure that data retrieval and presentation is seamless or at least appearing that way to the end user. This will require a close relationship with the subject matter experts, the end user and the client to ensure expectations are met.

Testing

Testing is one area that an EPM must focus upon to ensure the integrity of the system as well as system performance. Various types of testing must be completed on every eBusiness implementation. Success is dependant upon validating all of the work that gets done by various developers and technical teams. The project manager from the start must consider the types of testing that will validate the project and become the measurement of success.

Consideration should be given to what type of testing methodology will be used, what type of e-testing tools will be used and how will we analyze the results when complete? Other EPM considerations; should any or all of this be outsourced? Would it be more cost effective to outsource? These are all considerations that the EPM must keep in mind before putting the final test plan in place.

What should you look out for as the project manager if you decide to complete the testing in house?

Although this does not occur often and many times it is not intentional but there may be some bias when testing is done in house. For instance in smaller project teams the luxury of having an independent testing team may not be possible. The same people who built the system may be the only ones skilled enough to test it. Enter the EPM, who will make sure that the system is correctly tested and validated.

This section will discuss several levels of tests required for a User Acceptance Test Plan and the procedures for developing a User Acceptance Test Plan. To fulfill this task an EPM will need to understand the nuances of the following:

- Different levels of testing required for user acceptance testing for eBusiness
- Assessing the test type for a test plan
- Developing a user acceptance test plan

Different levels of testing required in a user acceptance test plan.

The possible levels of testing required in a User Acceptance Test Plan are:

- New system test
- Regression test
- Limited test

New System Test

The new system test occurs when the application to be tested is entirely new (is not an enhancement or system upgrade to an existing product).
Purpose: To ensure that the system meets all specified objectives and that all requirements are included in the new application.

Regression Test

The regression test occurs when the amount of change in an existing application requires a full system retest.
Purpose: To ensure that the "entire" system is working correctly, that application changes have not changed their existing functionality, and that the new development work meets all requirements.

Limited Test

The limited test occurs when the amount of change in an existing application requires only change-specific testing. When a full system test is not required, there are three types of limited testing: form-based testing, business process testing, and report testing.

Form-Based Testing

The purpose of this testing is to ensure that::

111

- Individual application forms are performing correctly
- All required fields and buttons exist on the application forms
- The flow of information and data entry is logical and correct based upon the application 's business requirements
- A new form added to an existing application is functioning according to specifications

Business Process Testing

The purpose of this testing is to ensure that all business related functions and processes are supported by the application, that the flow of data and screens is logical for each business process, and that security and access requirements are met; and to test the application with "real" scenarios, all batch processes, for performance related problems, and the application's interfaces.

Report Testing

The purpose of this testing is to ensure that the new report meets its requirements, the data extracted for the new report is correct, the report format is correct and logical, and the print process is working correctly.

Material that may require review includes: requirement documentation, system design documentation, technical specifications, contractor test plans, and contractor test results.

Assessing the Test Type for a Test Plan

The criterion for determining the degree and type of testing that may be required is listed below. This section may be used to determine what test scripting will be required for a particular User Acceptance Test Plan. If the application changes to be tested fall in more than one criterion, multiple test script types may be required. Once the appropriate test types have been determined, complete the User Acceptance Test Template as described in the section below.

A) New Application - (not replacing an existing application)

The Application Manager developing the User Acceptance Test Plan should be as involved as possible in the design and reviews of the new application with development team. The User Acceptance Test Plan should be developed with communication from the project Manager and with as much information gathered through the system documentation as possible.

B) New Application - (replacing an existing application)

The Test Plan should be developed using the required aspects of the system that are being replaced. Test scripts for any enhancements to the new (replacement) system should be developed as if the application is a new application - based on information obtained from the requirements and design documents. Running a parallel test with the old system and comparing critical report results would be the optimal test scenario for any functionality that is to be duplicated in the new system.

C) Database Change - (no other change to the application, such as an upgrade from Oracle 7.3 to Oracle 8.05)

In this instance, running a parallel test with the system using the old and new database is advised. Comparing critical report results would be the optimal test scenario to ensure that the new application database and drivers are producing identical results to the old system. Performance testing should be included in these test criteria. A User Acceptance Test Plan should be created to manage the parallel test.

D) Application Enhancements

Form-based test scripts should be developed for any forms that are being added to the existing application to ensure that each of the new forms is behaving correctly and is meeting the requirements. Business process test scripts should be developed and tested to ensure that the new functionality is integrated with the existing application correctly and that no existing functionality has been lost in the enhancement process. Test scripts should be developed based upon the specific requirements for any new reports. In this instance, unchanged forms may not require testing. For example, if they are not involved in any business processes that have been changed.

E) Infrastructure Change - (The application is not changing, but a test is required to port it to a new environment or a new server or the application be being utilized for a new division or purpose.)

In this instance, a full regression test of the form-based test scripts and all business processes should be repeated, because new environments can create unexpected problems for an existing application.

Developing a user acceptance test plan

To develop a User Acceptance Test Plan for a specific application, the Application Manager must first plan the tests based upon the initial application project documentation. By researching the purpose of the development work and the degree to which the development work will affect the rest of the application, the test planner can then begin to create the scripts for testing. (A test script includes the individual test steps to be executed in order to verify an application is working as expected.) The development of a User Acceptance Test Plan involves four steps which are described below and examples can be found in the reference section at the end of the book:

1. Developing the procedures and instructions for testing
2. Developing the necessary test scripts
3. Reporting any defects
4. Retesting any fixes

Developing the procedures and instructions for testing

This involves developing the necessary test scripts, and executing the necessary test scripts. The User Acceptance Test Plan Template is an independent document, which is intended to serve as a tool for the creation of an application-specific User Acceptance Test Plan.

Developing the necessary test scripts

This includes creation and customization of the instruction set? By including instructions for components assessed as necessary. Delete the unnecessary instructions from the template. Modify the appropriate application specific information and resources. For the EPM these tasks should be completed by the testing organization. In some companies, the organization may not be large enough to support a dedicated testing team. The EPM must understand these needs in advance of testing so that other arrangements can be made. For instance, third party testing companies are good source for impartial testing capabilities and resources. The down side of this approach is that the EPM will have to manage those resources.

Reporting defects

Listed below is an overview of defect tracking guidelines to use when reporting defects. You can create your own form and tailor it to your specific needs or modify an existing form. The EPM should make these available to everyone on the test team. As a reference prior to production and post production, tracking defects in a structured reporting format will save time and effort when problems occur in the future.

- It is critical to complete all fields on this form in order to assist developers in tracking and correcting the problem.
- A Defect Tracking Form must be completed at the time that a problem is found in order to assure that all details are documented correctly.
- Attach any available screen shots or report examples to the Defect Tracking form.
- Defect Tracking forms will be submitted *<insert how often forms should be submitted>* to *<Insert Application Manager info here>*, in order to implement fixes as quickly as possible.
- A service affecting defect (a problem, which inhibits further testing,) must be reported immediately to *<Insert Application Manager info here>*, in addition to completing a Defect Tracking Form.
- The *<Vendor or Application Manager>* will assign a defect number to each defect for tracking purposes.

An example of the defect tracking form can be found in the reference section at the end of this book.

Retesting any fixes

All the steps described above are applicable to any changes /fixes applied to the modules in the project.

The test scripts created for a new system, for regression testing or for limited testing should be recycled with applicable changes applied throughout the application's life. By editing existing test scripts the User Acceptance Test Planning process can save time and money, as well as maintain the test quality, as key requirement information will be reused.

Chapter Summary

In this chapter we have examined the eBusiness system implementation issues. The chapter covers the important features of the eBusiness architecture and also gives a clear picture of different testing standards and also provides the reader with ready to use examples. Knowing what to look for is an important skill that the eBusiness project manager must learn. Many times this knowledge comes from experience working on similar projects. In order to get that experience the EPM may have to learn from the failures of their own experience or from others.

The key take away form this chapter for you is to understand that the EPM's role is a program not a project. Continuous learning, improvement and refining are the critical success factors for an EPM.

Chapter 10

Securing Your eBusiness Application

Chapter Preview

Doing business online provides many benefits. Access to customers; lower transaction costs; and increased profitability, just to name a few. Companies, large and small, all across the world are seeking to "cash in" on the opportunity to use Internet technologies in the operation of their enterprises.

Though the benefits of conducting business online are clear, there are still risks to doing business in cyberspace. One such risk is the possibility of having someone steal information from your eBusiness without you knowing it, defacing your company's web site, or even worse, impersonating your eBusiness in order to defraud your customers. These scenarios may seem unbelievable, but companies are falling victim to these and various other types of Internet security attacks.

One reason for the trend is the lack of development of a comprehensive Internet security strategy at the time the eBusiness application was being developed. When managing the implementation of an eBusiness application, the project manager is responsible for making sure that the eBusiness application is "safe and secure". Unfortunately, many project managers lack the experience, background and knowledge to be able to lead their project teams through the process of securing the eBusiness application. As a result, many eBusiness applications fall victim to hacker attacks that have put them out of business or damaged their credibility.

This chapter will discuss the process involved in securing your eBusiness application. We will also cover the different types of security threats that your eBusiness may be exposed to and security strategies and techniques that can be used to protect an eBusiness application from hackers. The content in this chapter assumes that the security of the eBusiness application will be designed, implemented, and maintained by individuals within your organization. However, it can be also used as a benchmark for working with an outside managed security service provider to develop and implement an Internet security strategy.

Internet security is often discussed, but rarely understood. In order to get a better appreciation regarding the value of eBusiness security, an EPM must understand what the term, "Internet security," truly means. From a high level, Internet security is often defined as a system design that will allow the "good guys" in and keep the "bad guys" out.

From a more detailed perspective, Internet security should be defined as a strategy that uses technology, software, and processes to ensure that access to an eBusiness application's records is provided only to authorized individuals based on their "need to know." This definition of Internet security is more accurate, because it focuses on the protection of the most valuable component of an eBusiness application: its data and records.

As the Internet has gained popularity in the business community, the number of security attacks have risen. According to studies done by the Federal Bureau of Investigation (FBI) and other market research firms, the growth in "cybercrime" is becoming more and more prevalent. One reason for the increase in attacks is the availability of easy to use security penetration tools. For example, there are several password-cracking tools available on the Internet that can be acquired for a small nominal fee.

Another reason for the increase in Internet attacks is the amount of money and assets that can be accessed. For example, the average bank robbery nets a criminal around $9,000, while the average retail theft brings in less than $1,000. In comparison, a cybercrime can bring in millions of dollars with just the click of a button. Even worse, identifying the perpetrator of an Internet attack is much more difficult, so the chances of getting away with the crime is better than other crimes—no photos, no fingerprints.

In addition to understanding the definition and value of securing your eBusiness application, you must also steer clear of the misconceptions surrounding Internet security. For example, one misconception is that an eBusiness network is secure when a firewall has been installed. This is not true. Although having a firewall is a good start toward securing an eBusiness application, it will not protect an eBusiness application from all security threats.

Another misconception regarding Internet security is that most security attacks originate from outside of the organization. This is also untrue. Though attacks do originate from "unfriendly" sources, according to a survey conducted by the FBI and other security research firms, over

80 percent of reported security violations were orchestrated by individuals from within the organization.

The Law and Internet Security

The legal community is becoming very active in the eBusiness marketplace. They are working to set guidelines and benchmarks for how eBusinesses interact with their customers and how information is protected and secured. For example, in the United States, the Federal Trade Commission (FTC) has setup a Web site (www.ftc.gov) where consumers can file complaints against an eBusiness.

The development of e-centric legal and regulatory guidelines is not limited to the United States, countries all around the world are getting involved. For example, in Australia the government has implemented an eBusiness best practices model which provides guidelines on how companies in Australia are to conduct business online.

Having a basic understanding of the legal issues surrounding the eBusiness marketplace and the organizations that are involved in the development of the guidelines is important. It will provide you with information that may allow you and your Internet security team to develop strategies that will reduce the level of legal liability that your organization may be exposed to. It will also provide you with information that can assist in building a more secure eBusiness application.

In addition to understanding the laws in your home country, you should also gather information regarding the applicability of laws and standards in the countries where the users of the application are located. For example, if an eBusiness based in the United States has customers that are located in an European Union member country, such as France, the Internet security environment will have to comply with the European Union (EU) Directive on Data Protection.

Types of Security Threats

When securing an eBusiness application, one of the most important tasks is to identify the security threats that the eBusiness application may be exposed to and the level of impact that the Internet security threat could have on the eBusiness application and its stakeholders. The most common security threats for an eBusiness application are:

- Denial of service attack
- Web spoofing
- Trojan horse applications
- Social engineering
- Viruses

Denial of Service

A denial of service attack occurs when a hacker attempts to disable an eBusiness application by sending a stream of large data packets to a Web server for an extended period of time. The intent of the attack is to create a bottleneck on the Web server so that end-user transactions cannot be processed. A denial of service attack can be compared to when a prank caller continuously dials a phone number and hangs up every time a person answers the phone. As a result of the prankster's calls, the receiver of the calls may not be able to do some of the other tasks that are more important, such as taking calls from individuals that are truly interested in holding a conversation.

The most common type of denial of service attack is the distributed denial of service attack (DDOS). In this instance, a hacker takes over multiple computers and uses them to send data packets to the target Web server. See Figure 10.1

Figure 10.1

Distributed Denial of Services Attack

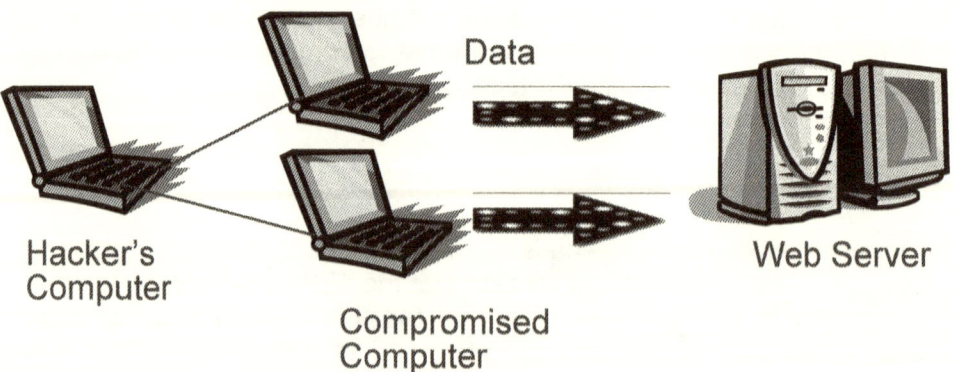

Data

Hacker's
Computer

Compromised
Computer

Web Server

Web Spoofing

Web spoofing occurs when a hacker creates a convincing, but false, copy of a Web site. The false Web site looks just like the real one; it has all the same pages and links. However, the attacker controls the false Web site, and all network traffic between the end user's browser and the Web goes through the hacker's computer. See Figure 10.2.

Other types of spoofing include Transmission Control Protocol (TCP) and Domain Name Server (DNS) spoofing. TCP spoofing occurs when an Internet packet is sent with forged return addresses, while DNS spoofing occurs when a hacker forges a computer's network address onto a trusted network.

Trojan Horse Applications

A Trojan horse attack occurs when an application is unknowingly loaded onto a web server in order to perform unauthorized activities. For example, a Trojan horse application can be used to record every keystroke executed on a target computer and transmit the information to that attacker's host server. With the use of new programming languages, such as Java, the detection of a Trojan horse application is more difficult, because many of the newer programming languages contain micro programs called applets that load onto a computer without getting the end user's approval.

Social Engineering

A social engineering attack occurs when an individual falsifies his or her identification or some other key piece of information to gain access to critical security information. An example of a social engineering attack is when any unauthorized user of any application calls the help desk to obtain a password and ID for accessing any application by obtaining an authorized user's name and employee number.

Viruses

Viruses are the most widely used method for penetrating a security environment or disabling the functionality of an eBusiness application. Viruses typically enter an eBusiness application as either an executable program or by being embedded into a document. A virus has the following four characteristics:

1. *Made up of a set programming instructions.*
2. *Deliberately created.*
3. *Actively propagates.*
4. *Infects other programs.*

Viruses go through the following three phases: dormancy, propagation and explosion. During the dormancy phase, the virus does nothing. This phase is often used by the creator of the virus to analyze the activity and architecture of the eBusiness application to find out where the application's security weaknesses exist.

The propagation phase involves the process where the virus is replicating or copying itself into other programs and into other areas of the Internet application, while the explosion phase involves the process where the virus begins attacking the eBusiness application.

Figure 10.2

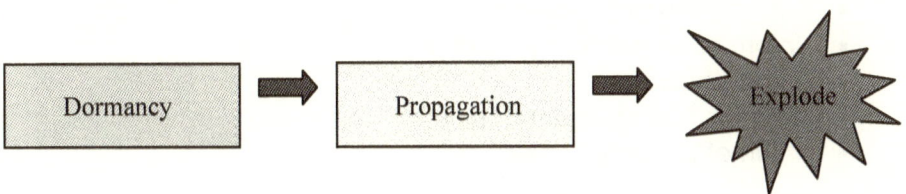

There are several different types of viruses. The ones that are most lethal to an eBusiness application are the stealth virus and the retro-virus. The stealth virus is designed to hide the changes made to computer files, while a retrovirus is designed to attack antivirus software.

Developing and Implementing an Internet security strategy

Securing an eBusiness application is a structured process. It involves performing the following tasks:

- Assembling a security team.
- Developing a security policy.
- Developing a security architecture blueprint
- Selecting a security technology.
- Customizing and testing the security environment.
- Training stakeholders on the use and maintenance of the Internet security environment.

Assembling a Security Team

Assembling an Internet security team is not easy. It requires a good understanding of multiple elements. The makeup of a project team will be driven by the scope of the Internet application and its risk management infrastructure, such as software and functionality. For example, an automotive company using a proprietary digital certificate solution that will be implemented globally to its employees and suppliers will require a team that has experience in digital-ID technology, as well as a good understanding of the online business processes that will be used by the end users.

In order to be effective, an Internet security team should include the following resources (See Figure 10.3):

- Network engineer
- Internet Security Administrator
- Information technology auditor
- Legal specialist
- Change Management Specialist/Trainer
- Internet Security Tester

The network engineer is responsible for designing the Internet security architecture. The person in this role should have a good understanding of the vulnerabilities that exist within each component of the eBusiness application's infrastructure (e.g. operating system, web server). They should also have a good understanding of security standards (e.g. x509) related to the eBusiness application's infrastructure and provide recommendations on how to best secure the eBusiness application.

The security administrator is responsible for maintaining the security environment once it is put in place. The person in this role should understand how to effectively manage user profiles and accounts. For example, if you decided to use a user ID-password approach to authenticate end users to the eBusiness application, the security administrator will need to know how to manage the password change control process and be able to identify when someone is trying to "breach" the eBusiness application's security environment.

The information technology auditor is responsible for identifying and providing recommendations on how to address access control risk through out the eBusiness application and ensure that end users are properly authenticated and authorized. The person in this role should have a deep background in business process control design and segregation of duties analysis.

The legal specialist is responsible for providing information and guidance regarding the interpretation of applicable laws that may impact how the eBusiness application will be secured. The person in this role should have a deep understanding of the different ways that countries around are world are using governmental and regulatory intervention to address Internet security concerns. They should also be able to help you build a system that can monitor the changes that take place in the Internet law arena.

The change management specialist/trainer is responsible for training the end users (e.g. business partners, internal staff) on how to interact with the eBusiness application without violating the organization's security strategy. They are also responsible for working with the stakeholder community to define or redesign organizational roles that may be needed in order to support the eBusiness application security environment. For example, a change management specialist may recommend that a new team-security emergency response team-be developed to respond to security breaches.

The security infrastructure testing specialist is responsible for testing the strength of the security environment. The person in this role should have a deep understanding of the different vulnerabilities that exist within infrastructure of the eBusiness application. They should also have some experience in using various hacker techniques.

Figure 10.3

Internet Security Team

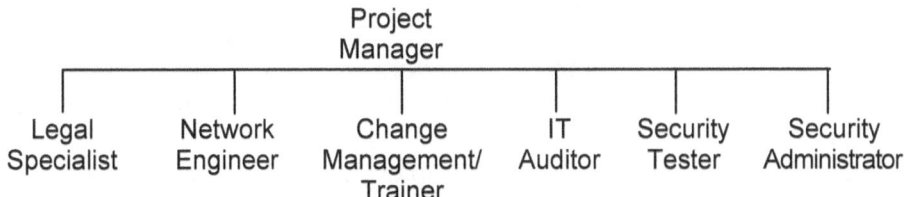

		Project Manager			
Legal Specialist	Network Engineer	Change Management/ Trainer	IT Auditor	Security Tester	Security Administrator

Developing a Security Policy

After assembling the Internet security team, the next step is to develop a security policy. A security policy is a document that provides information to users of an eBusiness application, such as employees and business partners, on how the application is to be used and secured. It also provides guidelines on how end-user access is to be managed (e.g. creation, maintenance, and distribution of end user passwords); how security violations will be identified, assessed, and resolved (e.g. disciplinary action); how security activities will be monitored; and who is authorized to use the eBusiness application.

When managing the development of the security policy, you should ensure that the different access points of the eBusiness application (internal, external, machine to machine) and the risks that may exist within each access point have been properly addressed. Also ensure that the security team has developed a security environment that uses the best security technologies and techniques to protect the most important resources that pass through or are stored on the eBusiness application.

From a technical perspective, the most important components that need be secured are:

- Web Server
- Operating System
- Database
- Data Transport Layer

Securing the Web server is important because it is usually the first point of contact by an end-user request and sometimes contains sensitive information. The Web server is also used to "serve-up" Web page content and manage Web pages. Ensuring that the operating system (OS2 WARP, Windows™, Linux™) is secure is important because it is the "heart" of the network platform. It coordinates the communication and activities between the different software applications that are a part of an Internet application. Many Internet security attacks are initiated against the operating system because it provides the greatest amount of control and opportunity for damaging an Internet application and any other systems that may be a part of the information technology framework.

The database system (e.g. Oracle™ and SQL™) must be protected because it is used to store sensitive information, like customer lists and credit card numbers, while the data transport layer of the eBusiness application (e.g. TCP/IP) must also be protected because it transports sensitive information in and out of the Internet application.

Developing a Security Architecture Blueprint

A security architecture blueprint is a document that provides a graphic depiction of how the end users, the technology architecture, and the security policy are integrated. The value of the deliverable is that it provides an overview of the entire security environment and allows for a more detailed comparison between what is in the security policy and what is actually being built. It also a serves as a high-level a road map for building the Internet security infrastructure.

Selecting a Security Technology

After the security architecture blue print is developed, the focus of the team should shift to selecting a technology enabler that can be used to implement the policy guidelines. There are several security technologies available in the marketplace, but selecting the "right" security technology can be a challenge. The best approach is to base the selection on technology's complexity, maintenance requirements, ability to address the exposures that have been identified, cost, vendor support, and penetration in the overall eBusiness marketplace. Using these criteria will ensure that the technology selected will be easy to use, accurately and effectively address the security exposures identified, have some market tested experience, will not "break" the project budget, and be backed by a company that has the "right" experience and background for providing support.

When examining the features and functions of a security technology, your assessment should focus on how the technology authenticates users, tracks end-user activity, and controls authorization to records and data once an end user is inside the eBusiness application. Other key features and functions that need be considered include the technology's administration related capabilities, such as deleting users and changing access rules, encryption standards compatibility and its recoverability capabilities in the event of a system crash.

When examining the ease of use and implementation, it is important to examine how many areas of knowledge are needed to maintain the security technology in comparison to other security technologies. Password-ID enabled security technology is easier and less time consuming than deploying a public key infrastructure or biometric security environment.

When considering the cost of the technology solution, the key area of focus should be whether the product is priced based on a per-seat basis or on a server basis. Understanding the pricing of the security application can save you several thousand dollars in unnecessary expenditures.

For example, if your organization has 5 servers and 100 users with the cost of a security solution being priced at $2,000 per installed server or $150 per user and all other measures being equal, it is cheaper to go with the installed-server pricing and forego the per-user pricing model ($10,000 vs. $15,000). To help qualify any differences in services, refer to the sample of a security selection framework in Figure 10.4. This template should be used to document information during the selection process. Fill out one template for each product being evaluated.

Figure 10.4

Security Selection Framework Template

Requirement Area	Requirement	Score (1-10)	Weight (.0-1.0)	Weighted Score (score x weight)
Features & Functionality	The security application must be able to positively authenticate end users and monitor their activity once inside the eBusiness application.	3	.4	1.2
Ease of Use	The security application must have easy-to-read screens, easy-to-locate buttons, and easy-to-remember commands.	4	.1	0.4
Cost	The cost of the security application must range from $1,000 to $3,000 and be priced based on the number of servers that the application is installed on.	3	.1	0.3
Market Penetration	There must be at least 500 or more customers in the eBusiness marketplace that are using the security application.	2	.2	0.4
Vendor Experience	The vendor must provide 24x7 support and implementation consulting services.	5	.2	1
Totals			1.0	3.3

When selecting an Internet security solution, it is very important to get a good idea of the environment you and your team are trying to secure. Understanding this concept is important because there is no one security tool that addresses all potential exposures.

Some of the security enablers/technologies available today include:

- Data encryption
- Digital certificates
- Load balancing
- Antivirus software
- Firewall
- Virtual Private Network
- Intrusion detection software
- Light Directory Access Protocol.

Data Encryption

Data encryption is a process that uses mathematical algorithms to scramble data while in transit over the Internet in order to prevent unauthorized individuals or systems from viewing or tampering with the data. There are different levels of encryption, such as 128 bit versus 40 bit. The theory is that the longer the algorithm stream, the harder it is to "break" the encryption structures and view the data. Some commonly used encryption standards are Secure Socket Layer (SSL) Secure Electronic Transaction (SET) and the Data Encryption Standard (DES).

Digital Certificates

Digital certificates are used in eBusiness to authenticate end users to an eBusiness application and protect information that is being sent over the Internet. Digital Certificates are part of a user-centric authentication scheme called public key infrastructure (PKI). A PKI system consists of two separate keys: a public key and a private key. The keys are used by individuals and organizations to send and receive encrypted electronic documents or gain access to an eBusiness application. An individual's or organization's public key easily accessible to anyone and is used to open documents or files that have been encrypted by the sender's private key. The private key is only accessible by the sender or key owner. Both public and private keys are issued by a Certificate Authority (CA)-an organization that takes legal responsibility for the identity of the individual or organization that is using the digital certificate. The certificate contains

Figure 10.5

Digital Certificate

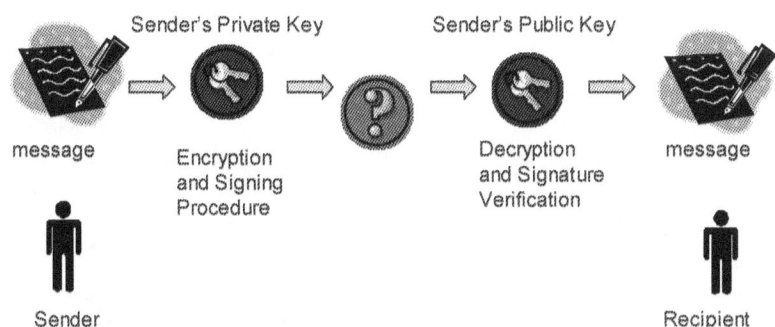

information about the owner, such as the public key, certificate validity dates and certifying authority's digital signature.

Load Balancing

A load balancing solution is software that can be used to redirect incoming data packets/traffic from one Web server to another Web server. Load balancing software is often used to defend against distributed denial of service attacks and unplanned spikes in incoming user traffic.

There are different types of load balancing techniques. One that is used quite often is geographic load balancing. This technique involves the redirection of incoming Internet traffic to Web servers in different geographic regions, like national and regional distribution. The value of geographically distributing Web servers is that it reduces the risk of an organization experiencing a natural disaster that could take out all of the Web server capacity at one time (also called single point of failure avoidance).

Antivirus Software

Antivirus software is the main technology enabler that is used to identify and remove viruses from a computer. Some of the key providers of antivirus software include McAfee and Symantec. The problem with antivirus software is that it is a reactive security defense. For example, when a new virus is discovered, the antivirus community begins working on developing a cure. Unfortunately, for those individuals and

131

organizations that were first to get the virus, nothing can be done once infected.

Firewall

A firewall is a software solution that is used to monitor incoming data packets. It is typically the first line of defense for protecting the network infrastructure of an eBusiness application. The value of firewall software is that it uses a rule-based approach to determine whether or not a data packet is acceptable. In addition to analyzing incoming data packets, a firewall can be used to hide the permanent Internet Protocol (IP) addresses of the network by using Network Address Translation (NAT) technology. In this way, hackers are not able to attack a server.

Virtual Private Network

A Virtual private network (VPN) is a technology that creates an encrypted dedicated connection between two computers over the Internet. A VPN uses two communication protocols: Point-to-Point Tunneling Protocol (PPTP) and IPSEC (Secure IP). PPTP was developed by a consortium that included Ascend Communications, 3Com/Primary Access, ECI Telematics, U.S. Robotics, and Microsoft. It is a nonproprietary networking technology that supports multi-protocol virtual private networks (VPN), enabling remote users to securely access corporate networks across the Internet, and complements firewalls by ensuring security of data exchanged between remote users and the corporate network. It supports various network configurations, such as analog phone lines or ISDN through the public switched telephone network (PSTN), frame relay, and x.25. It does not require any change to the network (e.g. readdressing of your network).

PPTP can be deployed in one of two ways. One way is to deploy PPTP drivers on the client machine and the server with all of the encryption being done on the client side and the decryption being done on the server side. Using this approach, no changes will need to be made by the ISP for a customer to implement this solution.

Another way to deploy the technology is to have the ISP install PPTP-capable dial platforms or front-end processors. Using this approach, any point to point (PPP) client that calls in, not just ones that understand PPTP, can establish a encrypted PPTP connection back to the corporation's PPTP server.

IPSEC, on the other hand, is set of IP extensions that are based on modern cryptographic technologies and offer strong data authentication and privacy guarantees by securing the network, rather than just the applications. For example, a normal IPv4 packet consists of headers and payload, both of which contain information of value to an attacker. The header contains source and destination IP addresses, which are required for routing, but may be spoofed or altered in what are known as "man-in-the-middle" attacks; the payload consists of information that may be confidential to a particular organization. IPSEC provides mechanisms to protect both header and payload data.

VPNs can be built over ATM, frame relay, and X.25 technologies. The current trend is for carriers to deploy IP-VPNs. The main reason is that "connectionless" IP-based VPNs offer enormous flexibility and ease of operation. Setting up an IP-VPN link between any two endpoints, anywhere in the world, merely requires configuring the communication at these endpoints. The IP network then automatically handles the routing of traffic between the endpoints. As there is no need to set up a connection first through all intermediate nodes or manually identify the best route, this greatly simplifies network planning and implementation. Because both the internal private corporate network and the new public VPN service are based on one common IP technology, the investment in equipment and training can be reduced significantly. Because the Internet is available almost anywhere in the world, in theory, any network could be simply interconnected to any other network using IP-VPN technology.

One important value of VPN technology is that it is based on the existing IP infrastructure used by the Internet and is protocol and platform independent. There are three types of VPNs: hardware-based VPN, firewall-based VPN, and standalone VPN application packages.

The hardware-based VPN systems are encrypting routers. They are secure and easy to use and provide the highest network throughput of all VPN systems, while the firewall-based VPN system is integrated into the firewall software and provides additional security for the backend network (e.g. Network Address Resolution).

A standalone VPN is ideal for situations where both endpoints of the VPN are not controlled by the same organization (typical for client support requirements or business partnerships) or when different firewalls and routers are implemented within the same organization.

Figure 10.6

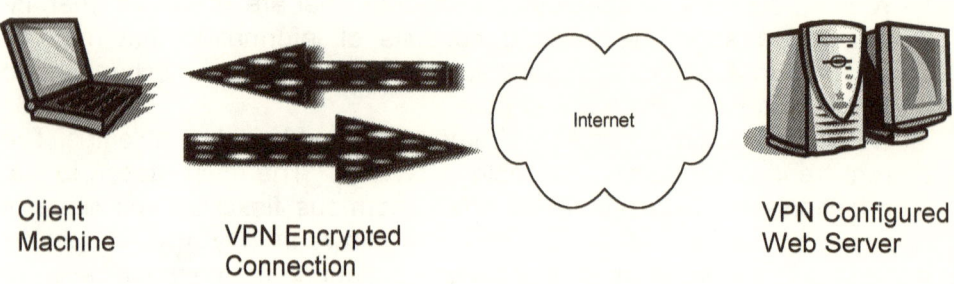

Client
Machine

VPN Encrypted
Connection

Internet

VPN Configured
Web Server

Intrusion Detection Software

Intrusion detection software is used to monitor activity and files on a network in order to identify and respond to hacker attacks. The software scans/monitors port activity and generates a log that can be used for analysis to determine if an attack is being waged against the network. There are two types of intrusion detection systems: network intrusion detection systems and host-based intrusion detection systems.

The network intrusion detection system (NIDS) analyzes network traffic for attacks that exploit the connections between computers and the data that can be accessed via a network connection, while a host-based intrusion detection system (HIDS) monitors specific files, logs, and registry settings on a single computer and can alert for any access modification, deletion, and copying of the monitored object. There are two types of detection techniques used by intrusion detection systems: sensors and agents.

Sensors are software programs that are used with network-based intrusion detection systems. It monitors incoming and outgoing traffic on a network. The software is placed in key locations around the perimeter of a network where it collects information regarding network traffic activities and reports on situations where the policy has been violated.

Agents, on the other hand, are software programs that are used with host-based network intrusion detection systems. They monitor changes made to specific files or logs on the host computer and report on situations where the policy that has been setup on the host computer has been violated.

The challenge with utilizing an intrusion detection system is properly configuring it to minimize the number of false-positive alarms—instances where there was no attack taking place, but the software set off an alarm. To help reduce false-positive alarms, it is helpful for the EPM to understand the two techniques used to identify policy violations on incoming and outgoing traffic: string matching and context analysis. The string matching technique examines traffic based on static components of the data packet, while the context analysis technique examines the static components of network traffic using a scenario-based approach.

For example, a company is interested in monitoring traffic to determine if its employees are violating company policy by purchasing airline tickets over the Internet rather than purchasing their airline tickets through the travel department. Under the string matching technique, if the intrusion detection system is configured to look for the strings "travel" or "airline tickets," all data packets with these terms embedded in them would generate an alert. This would create a lot of false positives because it would generate alerts when any of the company's employees would go on the Internet to view travel information.

Under the context analysis technique, a script can be written to find the previously described strings, but only when the data packets are outgoing and originate from an area of the company other than the travel department. With this approach, the number of alerts will be reduced, and the effectiveness of the detection process will be increased because the intrusion detection system will only focus on data that is probably originating from an area of the company that may require the support of the travel department.

Light Directory Access Protocol (LDAP)

The Lightweight Directory Access Protocol (LDAP) is a protocol for accessing online directory services. It runs directly over TCP and can be used to access a standalone LDAP directory service or to access a directory service that is back-ended by X.500. It is used on e-mail servers and as a tool for providing single sign-on access—a scenario where an end user is authenticated for access to multiple applications using one control point.

Using this technology saves time and reduces the likelihood of a security breach, because access is controlled at an enterprise-wide level and the users of the eBusiness application and other back-end systems are maintained using one access control list—unlike, some of today's current IT environments, where there is a separate user list for each application that is running on the network.

Imagine being able to add information about a new user through a single interface only once and immediately the user has a Unix account, an NT account, a mail address and aliases, membership in departmental mailing lists, access to a restricted Web server, and inclusion in job-specific restricted newsgroups. The user is also instantly included in the company's phone list, mail address book, and meeting calendar system. This is the promise of LDAP.

Customizing and Testing the Security Technology

Once the security technology has been selected, the next phase of work is to customize, implement, and test the security application. The customization and implementation of the security application should be done using the security policy that was developed. However, depending upon the applications being used to secure the eBusiness application, customization requirements may be different.

For example, let's say that a password/ID technique to authenticate end users and an intrusion detection software solution to monitor incoming traffic are chosen to secure an eBusiness application. When customizing the password-ID portion of the security environment, you and your Internet security team will need to decide on how long the password will be, whether or not it will be text only or will include both text and numeric characters, how the system will be used to reset passwords, and how the system will be configured to protect against a brute force attack. When customizing the intrusion detection software, the Internet security team will need to determine what type of data monitoring technique will be used (network based vs. host based), what technique will be used to monitor traffic that passes through the network (sensors vs. agents), and how the intrusion detection system will identify hacker attempts (string matching vs. context analysis).

Testing of the eBusiness application will involve the development of a test environment, such as test cases and the use of security testing tools that may help in identifying any technical or strategic exposures. Testing a security environment can be difficult because there are several ways that it can be exploited.

One of the best ways to test the security environment is to test each layer or component of the eBusiness application's technical architecture (network layer, communication layer, and application layer). For example, when testing the security of the communication layer, the goal should be to ensure that data that is either received or transmitted by the eBusiness application cannot be viewed or tampered with during transmission. One way of doing this is to send an encrypted packet of data across the network and then attempt to view the data as it passes through the network using a packet sniffer.

Training End Users

Training end users on the use of the security environment is one of the final activities in securing your eBusiness application. It involves the development of training materials and structured education that guides end users through the process of activities, like requesting a password-ID, resetting passwords, and obtaining a public-private key.

The deliverable that should be produced as part of this process is a security training plan, which should answer the following questions:

- Who will be trained?
- What will be included in the scope of training?
- What is out of scope for training?
- When will training begin and end?
- Who will be responsible for developing the training content and conducting the training?
- How will training be delivered?
- How will training be updated and maintained?

Of the questions listed above, two of the toughest questions to answer are usually, "Who will be responsible for developing the training content and conducting the training?" and "Who should be trained?" The best answer to the first question is to use a team-based approach to training. For example, the development and delivery of the security training materials should be done by two individuals: one that understands how to utilize the security technology and one that understands how to communicate technical topics to the end-user community.

In regard to the second question, the training focus should first be on those individuals or groups of individuals that are directly impacted by the eBusiness application. For example, in an e-banking application deployment, the top three target groups for training should be the external users, such as customers and business partners; internal staff that support external users; and sales personnel that are responsible for selling and promoting the e-banking products to external users.

Chapter Summary

This chapter discussed the importance of securing an eBusiness application, law and Internet security issues, security advocacy organizations, types of security threats that an eBusiness can be exposed to, and security strategies and techniques that can be used to protect an eBusiness application from hackers. With this knowledge, you are now ready to proceed to Chapter 11, which will focus on how to manage the process of protecting the privacy of the individuals that use your organization's eBusiness application.

Chapter 11

Protecting Consumer Privacy

Chapter Preview

A major factor in competing in the digital economy is knowing your customer. This includes not only knowing what they purchase, but also things like when they purchase, what pages on the website they view; and where they are located (e.g. Michigan versus Texas). If your organization is a retail eBusiness, having information about the gender and age of the customer may also be of value.

Having a good understanding of the factors that drive an individual to visit your organization's website and make a purchase are crucial to ensuring long term profitability and good customer service.

Using the Internet to learn more about your customer's habits is easy to do, but it must be done in way that it does not invade or violate the privacy of the customer. The consequence of not protecting and respecting the privacy of the individuals that use your organization's eBusiness application can be very damaging.

In order to avoid the risks related to online privacy protection, you (the project manager) must be aware of how to ensure that your organization's eBusiness application is being designed, configured and operated in a way that ensures that privacy of all eBusiness application users is properly protected.

The purpose of this chapter is not to provide you with legal advice regarding online privacy. You will need to hire a lawyer for that. It is designed to give you a baseline understanding of the online privacy landscape and how to lead your project team through the process of designing a privacy strategy that is effective, actionable and manageable. It is also designed to provide you with information on organizations that you can go to for support (e.g. keeping up with new privacy threats).

Understanding Importance of Consumer Privacy

Privacy is a hot issue. It is not a new concept because sensitive information about individuals has been collected for decades (e.g. public records, phone books). Some reasons why privacy in the digital economy is more of a concern are that information can be inconspicuously gathered by anyone over the Internet and may be used in a way that is not acceptable by the information-owner. For instance, an eBusiness can

139

gather information regarding an individual's income, gender and medical history for a small nominal fee and then share it with one of its business partners. The impact to the information-owner is that they may receive unwanted solicitations. Even worse, the information may be used to commit fraud (e.g. identify theft).

Another reason why privacy is at the top of the list of concerns in the digital economy is the increase in litigation where eBusinesses engaged in the collection and sale of sensitive information have violated the privacy of thousands of Internet users. For example, in January 2000 DoubleClick, an online advertising and marketing firm, was sued for tracking the web surfing activities of individuals and sharing that information with its business partners and clients without the explicit permission of the Internet user. The company used a browser-based technology called *cookies* which captures information things like what sites an individual has visited and the IP address-a computer based location identifier-of the individual's computer. Other high profile cases that have involved the violation of an individual's online privacy rights include RealNetworks and the Federal Bureau of Investigation (FBI). In the RealNetoworks case, the company was chastised by the eBusiness marketplace for using RealJukebox, a software used for listening to online music, to track what CDs an individual was listening to without the individual's knowledge or permission. In the case involving the Federal Bureau of Investigation (FBI), the organization was criticized for using its *Carnivore* system to monitor email and instant messages that cross the Internet without the permission of neither the sender or the receiver of the email message.

Having an understanding of online privacy protection issues is important because it may impact the level of legal and financial risk that your organization's eBusiness may be exposed to. It is also necessary to understand this issue in order to develop a strategy that will ensure that your organization's eBusiness application adheres to the online protection privacy guidelines that have been put in place by legal and regulatory agencies.

Security versus Privacy

Privacy and security are often used synonymously. Unfortunately, privacy is not the same as security. The differences between the two terms lie in who is the decision maker and what processes of the eBusiness are being impacted.

From a decision-making perspective, online security is controlled and managed on a group level, while online privacy protection is controlled and managed on an individual level. For example, the development of an organization's security policy is based on how it wants to protect access to its internal assets (e.g. data stored on web server and back end systems), while the protection of an individual's privacy is based on what information the individual is willing to provide and how they want information about them used.

From a process impact perspective, online security is focused on how the information is secured when its being collected and transmitted over the Internet or stored on an Internet-connected database, while online privacy is focused more on getting the proper authority from the individual regarding when and where their information collected as well as who their information is shared with.

Understanding the differences between the two terms is important because it will impact the functionality and features of the eBusiness application; the scope of work that the eBusiness project will involve; the size and uniqueness of the project team and the type and complexity of the online privacy protection and security strategies that are designed and deployed.

Privacy Trends

The online privacy protection movement is experiencing radical change. Some of the changes include the increased involvement of government in the development of online privacy guidelines; the development of privacy protection and privacy exploitation technologies; and the development of new organizational roles within eBusinesses that are responsible for the development and management of the organization's online privacy initiatives.

Government Intervention

All levels of government (state, local and federal) are working to develop and implement of laws that will provide eBusinesses with guidelines regarding online privacy protection. The focus of most of the proposed legislation is on what information can be collected; getting the approval from the information-owner on how their information is shared and what it is used for; and providing the information-owner with the opportunity to review, at anytime, the information that has been collected about them.

Some of the US based governmental organizations that are actively involved in formulating online privacy guidelines include the US Congress and the Federal Trade Commission. On an international level, governmental institutions in Asia-Pacific, Europe and South America are also working on developing online privacy guidelines.

The number of proposed guidelines is growing rapidly. Also, the guidelines are being designed to industry specific online privacy issues. For example, in the US, the Financial Information Privacy Act was implemented by the Federal Reserve Bank to address issues related to online banking, while the Health Insurance Portability and Accountability Act (HIPAA) was implemented by the Department of Human Health Services to address privacy issues in the health care marketplace. As the Internet continues to mature, there will be additional guidelines put in place. As the project manager it is important to be aware of the changes in the legal landscape.

Below is a sample list of some of the legislation that US governmental and industry organizations have passed into law.

Sponsoring Organization	Name of Legislation	Description
United States Congress	Online Privacy Protection Act	Requires privacy disclosures; allows consumers to access their personal information.
Federal Trade Commission	Children Online Privacy Protection Act (COPPA)	Focused on protecting children while they are online by requiring web sites that target children under the age of 13 to disclose their privacy policy and use technologies that will positively identify web surfers.
Department of Human Health Services	Health Insurance Portability and Accountability Act (HIPAA)	Focused on the protection of an individual's medical information; allows patient access to personal medical information.

Privacy Protection/Exploitation Technologies

Another trend in the online privacy market place is the development of new privacy-based technologies. Privacy based technologies and services are being developed for eBusinesses and web surfers. On the eBusiness side, organizations are using technologies to track the web surfing activities of the individuals that visit their website. Two technologies that are often used are cookies and web bugs. Cookies are applications within the web browser that are used to track an individual's web surfing habits. Examples of the type of things that are tracked by cookies include web browser type, link/web page visited and time a link/web page was viewed.

On the web surfer's side, there are online services available that allow an individual to surf the Internet without leaving an audit trail-anonymous browsing. The value of the service is that it protects a web surfer's privacy by allowing them to retrieve information from a remote web site using an encrypted connection (e.g. SSL). Other technologies that are being developed for web surfers are solutions that block cookies and Java Script from loading onto the web surfer's computer. There are even solutions that will determine what the purpose of a cookie is before loaded onto a web surfer's computer. This technique helps the web surfer make a more informed decision on whether or not a cookie should be accepted.

New Organizational Roles

With online privacy becoming more of an issue in the eBusiness market place, some organizations have created roles (e.g. chief privacy officer) within their organizations to proactively deal with online privacy risks. A role that has gotten a lot of attention is the Chief Privacy Officer or CPO position.

Some of the responsibilities for a person in this position include monitoring the organization's adherence to its online privacy policy; responding to privacy complaints; assisting in the development of privacy initiatives and educating stakeholders (e.g. employees, customers and business partners) on changes in the privacy landscape that may affect them.

The value of the position is that it is a quasi-independent role that has a more proactive approach to privacy protection and can create a positive public image for an organization that is heavily involved in eBusiness related activities.

The person in the CPO position often reports to the head of legal counsel. In the case of more flat organizations, the CPO may report directly to the Chief Executive Officer (CEO). As the Internet becomes more integrated into day-to-day business activities, companies conducting business online will need to establish a CPO position.

Privacy Advocacy Organizations

As interest in online privacy has grown so has the number of organizations that are designed to monitor and develop online privacy standards and activities.

There are two types of privacy advocacy organizations: privacy rating organizations and privacy standards organizations. The privacy rating organizations (e.g. BBBOnline.com) are involved in examining the privacy guidelines of an eBusiness and performing an assessment of how closely the eBusiness follows its own privacy policy. In the event an eBusiness is found to be in violation of a generally accepted privacy practice, it is given a low rating/score by the privacy rating organization.

In contrast, privacy standards organizations (e.g. Electronic Privacy Institute) work with industry and governmental agencies to develop standards that can be used to protect the privacy of a web surfer while they are online.

Developing an Online Privacy Strategy

Ensuring online privacy is not about writing a three or four line privacy statement and posting it on a website. Its more about understanding what to include in the statement; how to track and monitor your organization's adherence to its privacy strategy; what business processes are being impacted; what information is being collected and who the information is being shared with.

The key thing to remember is that as the project manager, your role will be to manage the development process and ensure that the final product is effective and addresses the key privacy risks that are associated with running your organization's eBusiness. You should not be tasked with doing any legal research or provide legal advice. You are, however, responsible for making sure that your project team is aware of issues surrounding online privacy issues and that they are addressing these issues through out the design and implementation of the eBusiness application.

In order to successfully *manage* the design of your organization's privacy strategy you will need to task your project team with performing the following activities:

- Developing an online privacy mission statement.
- Performing an online privacy assessment.
- Selecting an online privacy approach.
- Documenting and distributing privacy strategy to the stakeholder community.

The privacy mission statement is a one to three paragraph statement that explains your organization's privacy goals. It is often short and straight to the point. The development of this statement is often done with the assistance of professionals with experience in dealing with privacy related issues (e.g. CPO).

The performance of an online privacy assessment involves having members of your project team conduct a thorough analysis of the cyber laws that may have an impact on your organization's eBusiness strategy and operations. It is important to perform this activity because online privacy guidelines differ from country to country. The value of the privacy assessment is that it ensures that privacy is considered through out the eBusiness application development process. Examples of tasks performed during an online privacy assessment include identifying where the eBusinesses web traffic will be coming from (e.g. France, United States, Asia); getting an idea of the type of laws that are in place in these areas to determine what the privacy requirements are (e.g. collection and sharing of personally identifiable information, data security, training and communication); and calculating the financial and environmental impact that a privacy violation can have on your organization's eBusiness. Some of the other tasks that should be performed by your project team during the performance of an online privacy impact assessment include examining what processes within the eBusiness are integrated into the privacy side of your organization's eBusiness. For example, in a retail eBusiness the online payment and order management processes contain activities that have privacy protection overtones (e.g. collecting shipping information or gender information). Having a basic understanding of the integration points between the processes and online privacy environment is important because it will help you determine where potential exposures may exist within the eBusiness infrastructure.

Once your team completes the privacy impact assessment the next scope of work is to select an online privacy approach.

Selecting an online privacy approach involves taking the information that was collected during the impact assessment process and determining what approach will be the most effective at achieving the goals that are included in your organization's online privacy mission statement and the requirements that have been set by governmental and regulatory agencies.

An online privacy strategy can be simple or complex. The determination of how complex or simple the strategy will need to be is a function of the environment in which your organization's eBusiness will operate and the amount of resources (time, money, people, technology) that will be needed to implement the strategy.

For example, some eBusinesses have chosen to simply put a text based privacy statement on their web site, while other businesses have chosen to place a privacy policy on their web sites and also take part in an independent privacy audit.

The cost and time involved in designing and implementing the first strategy, a text based privacy statement, may only require access to the website and the expenditures for hiring a cyber law specialist, while the second strategy, a text based privacy statement with an independent privacy audit, may require not only access to the website and the cost of a cyber law specialist, but also systems that can perform privacy-adherence monitoring functions (e.g. identifying customers that have opted out of an email promotion program).

After a privacy strategy has been selected, the next scope of work for your project team is to document the strategy in a format that is easy to read and understand. Documenting the strategy is important because it provides a paper trail for impacted stakeholders to review if there is a disagreement that arises about how the organization is protecting the privacy of its online customers and business partners. As part of the documentation process, the document (often called a privacy policy) should include, at a minimum, when information is being collected; what information is being collected; how the collected information is being used; and who the collected information is being shared with (e.g. business partners); how a individual can protect their from abuse and how privacy complaints are handled (e.g. filing, resolution, contact person).

Chapter Summary

In this chapter we looked at online privacy protection and the responsibilities of the project manager in ensuring that their project teams are aware and responsive to privacy related issues. We also discussed how the growth in the concern regarding online privacy has impacted the eBusiness market place (e.g. establishment of new organizational roles and privacy watch-dog organizations) and what process that can be used to develop an online privacy strategy.

The goal of providing this information is not to provide you with legal advice because none of the authors of this book are lawyers, but to give you some sense of what is required of you as a project manager to ensure proper protection of information that is provided by users of your organization's eBusiness application.

Chapter 12

Deploying eBusiness Projects

Chapter Preview

This chapter will focus on how an EPM can refine the deployment strategy by using several methods, techniques, and communications. The deployment strategy should set a process in place that can ensure that the benefits of the solution can be sustained over time. To do so, an EPM should begin by reconciling the first production period data for evidence of possible performance and enhancement opportunities. Each production area has a place in the success of the deployment. Additionally, an EPM should understand that a business solution deployed today is a moving target and that enhancement, change, and evolution are factors that are fundamental to success.

To help an EPM organize the deployment strategy, this chapter will define the stages of a roll-out strategy, discuss factors involved in developing a support strategy, and explore some methods and strategies for training end users.

Preparation for the eBusiness Application Roll-Out Strategy

At this stage in an eBusiness project, an EPM should be asking how the application will be delivered to the user community. Many times, the deployment phase is where project management begins to feel that they have accomplished the task at hand and begin to sit back and relax their efforts. This is where many projects get into trouble. The sooner the EPM realizes that the project is only the three-fourths of the way through, the more successful the project will be. The EPM's work is not complete until the project is handed off to the incumbent postproduction owner. Good project managers understand that the project is long from being complete until it is formally handed off to the support team. Even then, it is up to the EPM to ensure that the support team has the training, documentation and expertise to support the application. When possible it is best to engage the support team as early in the project as possible to ensure a smooth transition.

The deployment phase of the project will bring new challenges to the project manager and to the team. Rolling out the eBusiness application to the user community will involve many of the concepts discussed in previous chapters as well as some additional strategies. For instance, the target market or audience should have already been well defined. However, there may be areas of concern that have not been accounted for. Examples of some deployment issues that can arise are listed in Figure 12.1.

Figure 12.1

Deployment Issues		
Software and Data Migration	Fine-tuning the Application	New Roles and Responsibilities
Management Tools	Troubleshooting Tools	Network Management
Change Management	Documentation	Increased Support Staffing

If the EPM uses the approach detailed throughout this book, he or she should have already integrated many of the tasks that will be incorporated into the roll-out plan from the requirements and design phase. With these elements in place, an EPM should now be able to focus on the delivery methods. Not unlike the other phases of the project, deployment must have a method, a plan, and a way to measure success. As indicated in Figure 12.2, the process for deployment may be outlined as follows: vision, detail, test, deploy, and enhance.

Figure 12.2

Deployment Process

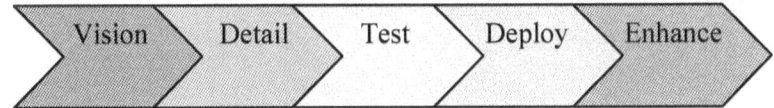

Vision

Deployment starts with vision and focus. This stage may include gathering background information on the target audience, mobilizing project teams to focus on the deployment task, and analyzing information already gathered. In the deployment phase, these tasks can be broken down into activities and deliverables. The EPM is responsible for making sure that each task is being executed.

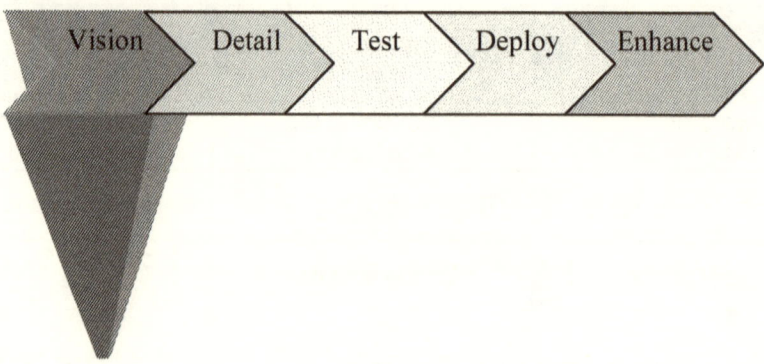

Vision Activities:

- Complete deployment team staffing
- Redefine communications plan
- Address final scope changes
- Confirm environment readiness
- Complete stress testing, staging, and data validation process

Vision Deliverables:

- Updated communications plan
- Final project plan
- Critical-issue resolutions implemented
- Clearly defined change control process

Detail

Using the project information, an EPM should next develop a detailed deployment process for management review. This process should identify what best practices will be employed and any risks and constraints. It should also refine the business case model in more detail.

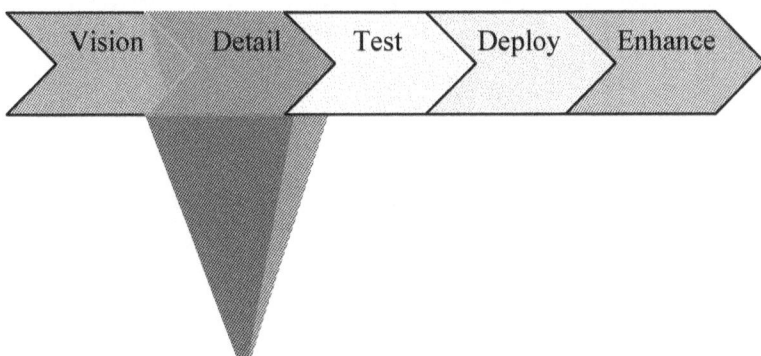

Detail Activities:

- Develop training formats and delivery methods
- Confirm and validate all custom development work
- Ensure that all core and support team members have been trained
- Conduct mock-up scenarios using real data
- Validate connectivity

Detail Deliverables:

- Support team training documented
- Functional specification review and validated
- Traceability matrix confirmed
- Validate end-to-end business process

Test

Testing itself is a project. The test team should have been involved at the high-level design phase of the project. Test scripts should have been written specifically for the project. A user acceptance team (UAT) should have been identified and be ready to begin the testing phase. An EPM should always monitor progress and findings during a UAT. A regression-testing plan will be key to developing successful code changes and fixes to manage the revision process. See chapter nine for a detailed test plan.

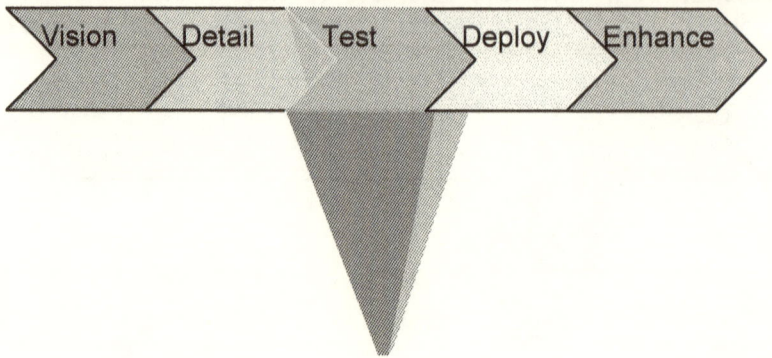

Test Activities:

- Identify the test team
- Develop or revise test scripts
- Identify user acceptance test team

Test Deliverables:

- Test plan
- Regression-testing management plan
- Test results

Deploy

Deployment strategies should be well defined in the project charter. As the project progresses changes and adjustments to the deployment strategy should be enabled and documented.

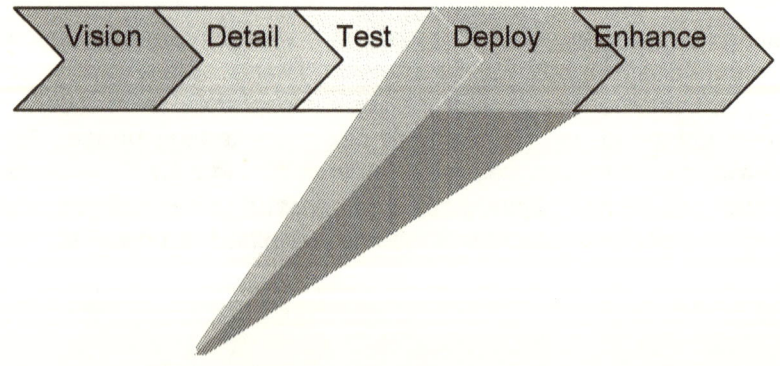

Deployment Activities:

- Implementation of the communication plan
- Go-live review
- Production hardware setup finalization
- Complete connectivity issues

Deliverables:

- Hardware check-out results
- Back-out plan
- Disaster recovery test and recover results and plan

Enhance

As with any product or service, continuous enhancements and improvement must be a part of the overall deployment strategy. Some incremental revisions should be anticipated as well as, longer term scheduled releases with more functionality. The planning process for enhancement should begin prior to deployment to account for any features or functionality that did not make the initial deployment.

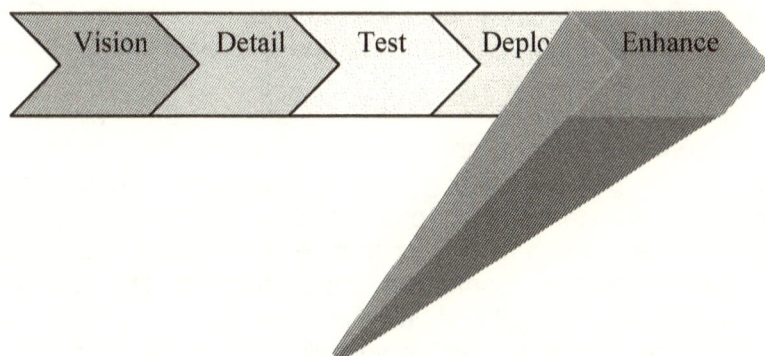

Enhance Activities:

- Distribute change control or enhancement process
- Solicit feedback from user community
- Evaluate results based on predefined measurements

Enhance Deliverables:

- Enhancement process document
- Updated system change repository
- Revised user, training, and administrative guides

Postproduction Support Strategy

To define the postproduction support strategy, an EPM will need to determine what support will be needed, identify what human resources will be used to provide that support, and determine the funding needed to implement that support strategy. One area of concern is the assignment of responsibility for the day to day maintenance of the site, usually assigned to the webmaster. The webmaster must be responsible for maintenance tasks that include such activities as Information updates (refreshes), hyperlink validity, navigation consistency and aesthetic consistency.

Determining The Support Needed

The people and the process combined are one of the three components for a successful post implementation strategy. Customer service begins with these components and the planning that goes into supporting the program.

The second component is customer service. One way to look at customer service is to think about personalization. Personalization is a process that changes the starting point of marketing. No longer does a company have to start with a product idea, design it, implement it, manufacture it, and cross its fingers that all of the sales and ad revenue will sell it! Now, the ideas come from the customers. The customer tells the company what they want and how they want it.

The concept of customer-driven products and services has made selling the product possible, easier, and cheaper by the advent of technology, but it is not just technology. It is a new way of looking at an operation's purpose. This should excite the customer service department.

The third component is the help desk support team. If the help desk team is engaged or created early in the conceptual phases, it will help ensure that they will be ready when needed. The help desk team should be an integral part of the testing strategy, be involved with the acceptance testing and the development of response procedures. They should also be involved with and contribute to the support escalation plan.

When an EPM begins thinking about what issues and problems the users could face once the project goes live, he or she should think about the unthinkable! Has the application been tested on various versions of browsers? Has it been tested at different connection speeds and portals?

For example, when one application was developed, all the users had high-speed, high-bandwidth access on the secure side of a firewall. Shortly after phase one going live, most of the users were using either VPN or dial-up connections from remote locations outside the firewall. The result was a very slow response time and frustrated users. For reasons such as this one, an EPM should test every possible scenario to see how the application performs.

An application stress test can be conducted by either using a commercially available tool designed for stress testing or by having an independent third party complete the entire testing phase of the project. It is not necessarily a bad idea to have this part of the project looked at by an unbiased set of eyes. In fact, this approach can pay big dividends by identifying errors that may have been over looked by a team that was too close to the project to see them. Third-party testing can also free up resources that can be used on other segments of the project.

Another component in any post implementation strategy is communication, with the application development team and the user community. Communication must be delivered and received from above, below, and laterally. It must be asynchronous, meaning that both parties must engage each other to facilitate constant improvements and feedback.

Identifying the Human Resources Needed to Provide Support

Now that the help desk is established, how many resources will be working on responses and support? Is the help desk a 24/7 operation? How will support be delivered to various clients throughout the world? What languages will the help desk support?

Will support be a three-tiered approach with level one being the help desk, level two being the technical support team on call, and level three being a contracted support program maintained by subject matter experts at the company that the software was purchased from or your own development team? Will the use of e-mail support be part of the response mechanics?

All of these are valid questions that can only be answered on a case-by-case approach. A one-size-fits-many approach seldom works effectively. Customers have come to expect customized support, service, and response.

Determining the Funding Needed to Implement the Support Strategy

What will it really cost to implement a customer service center or an automated customer support software to manage the support of an eBusiness solution? Will this strategy be able to implement the business policies and processes?

When the estimate begins to take shape, one approach to help determine the costs is to define the "levels" of service that will be offered. A sample service level offering is outlined below. Once this is defined, it becomes much easier to identify the costs and resources needed with each level.

Platinum Support
    Critical - Critical business impact
    One-hour target response time
    Business process is severely impacted with no procedural capabilities.

Gold Support
    Serious - Serious business impact
    Two-hour target response time
    Business process is affected, but there is no immediate business impact.

Silver Support
    Normal - Medium business impact
    One business day
    May experience some loss of functionality, but no business impact.

Bronze Support
    Wish List - Request for services
    Two business days
    Request for an enhancement or information.

Training End Users

When developing a plan for training end users, an EPM will first need to identify the user training audience. With this knowledge, he or she can then begin to evaluate the various delivery methods and strategies that can be used for end-user training and then develop a training plan.

Evaluating Various End-User Training Methods and Strategies

Before developing a training plan or curriculum, your target audience must be identified. (This question should have been answered early on in the requirements phase of the project.) Once determined, the various options for training delivery can be evaluated. Some of the delivery methods and strategies used for end-user training include:

- Just-in-time training may be used to ensure that newly acquired skills and knowledge are put to immediate use.
- Measurement of the training should be evaluated in two areas: reaction and learning.
- All training must be placed in an environment that is conducive to learning.
- If an instructor-led program is used, the class size should be limited in order to optimize contact with the instructor and to allow for one-on-one consultations.
- A "train-the-trainer" approach should be used when appropriate to ensure that the training remains consistent over time.
- The training program should incorporate various media, such as computers, overheads, printed course manuals, whiteboards, online documentation, slides, and even Web-based meetings tools.
- Various methods of presentation should be used, such as demonstration, discussion, lecture, discovery, reading, and tutorial, because people learn by different methods.

The training will measure the overall success of the program. When the audience is primarily adults, the course should apply the basic concepts of adult learning, such as:

- Self-directing learning techniques - adults prefer to discover new things on their own accord.
- Experience - they come to class with both positive and negative experiences.

- Instant gratification - "what the system will do for them now and why it is better than the old way." Be prepared to address this.
- Problem-solving - many adults like to work on problems and exercises.
- Self-esteem - they can get their feelings hurt easily.
- Learning - adults learn by doing, but understand they also learn at different rates and by different means.
- Continuous feedback - they need to know how they are doing.

When considering training plan methods and strategies, it is important to keep in mind that the major component of a successful implementation is the people. After human resources, the next major component is the training method of the program. When being trained on a new system, the trainees will need access to and continued use of the system to retain the skills and knowledge they acquire. The training program should be as short as needed for the target audience and scheduled as close as possible to the actual time that the system will be employed. This will keep their skills fresh and make the transition easier.

Developing a Training Plan

There are many options available today for delivering training to those who need it. For example, remote Web-based training is coming of age as bandwidth becomes less costly. Computer-based training (or CBT) is still a viable option, and traditional classroom training still remains one the best delivery methods. Coaching and role-based training can be a good selection if the resources are available for delivering this method in a cost-effective way. One of the best uses of the Internet is the ability to reach many people in various geographically dispersed locations at the same time.

As technology advances, the Internet is becoming more cost effective and user accepted. There are several companies that can be used to deliver these types of training programs to your audience. The use of web based meeting applications like NetMeeting and Webex are especially suited for this type of delivery and role-based training. Role Based training takes a subset of the overall system training and focuses on a specific area or role. This facilitates breaking up the overall training program to deliver more specific, detailed instructions to those who may only need to use a specific module or function of the system.

Online help and interactive assistance should be incorporated into the program and should be intuitive enough for any user to be able access and use it as the first stop for help and assistance. Mouse-over and links are also very effective ways to give real-time help.

Chapter Summary

This chapter defined the stages of a roll-out strategy, discussed the factors involved in developing a support strategy, and explored some methods and strategies for training end users.

As a project manager, it is important to understand that the project is not over until an official transition to the postproduction team has taken place. "Many eBusiness solutions never really end. They just continue to evolve." These are words that an EPM must live by.

Reference

Reference Section Examples

Form-Based Test Scripts

The purpose of testing the individual forms and the user interface is to assure that all of the menus and graphical interface buttons, pull down lists, scrolling lists, and check boxes are performing correctly. It is important to perform each interface for each module/screen as not all modules have the same buttons or menus.

A procedure for testing the User Interface may look like this:

1. Develop user Acceptance Test Script as shown below. Develop script for each module screen listed in the "List of Modules" spreadsheet.
2. Mark at the top of each form-based test script which module/screen you are testing.
3. Complete the test environment section at the top of each test script.
4. Login as System Administrator (the highest security level in order to have access to all screens).
5. For each step in the test script initial and date the successful tests.
6. If the test results are not as you expect, or a comment or clarification on a screen or process is required, please complete a defect form for that test step. (Do not initial and date unsuccessful test cases).
7. Please return a copy of all completed Test Scripts and Defect Forms as per the rules and timeframes described under that Section.

Figure 1: List of Modules Example

Module ID	Module name
CRS0000	Main Menu
CRS1020	Enter Customer Login Scenario
CRS1080	Enter an Authorization Scenario
CRS1050	Enter Customer Data Entry Scenario

Figure 2: Form-Based Test Script Example

Form-Based Testing Component	Date	Initials
1. Are all fonts, colors, shading, and toolbars consistent with standards and project guidelines?		
2. Is the online help mo 2dule available?		
3. Are all date formats correct? (DD-MON-YYYY) Are the date edits being correctly applied? Are dates greater than 2000 accepted?		
4. Does all text wrap when displayed in the text editor?		
5. Is there a scroll bar on every applicable block?		

Business Process Test Scripts

How to Customize the Business Process Test Scripts:

- Develop Business Process Test Scripts as shown in the figure 3. Develop test scripts for each user level listed in the User Security Matrix.
- Mark at the top of each Business Process Test Script which user level you are testing.
- Complete the test environment information as required at the top of each Business Process test script.
- Login as the appropriate user level.
- For each step in the test script initial and date the successful tests.
- If the test results are not as you expect please complete a defect form for that test step. (Do not initial and date unsuccessful test cases).
- Please return a copy of all completed Test Scripts and Defect Forms as per the rules and timeframes described under that Section.

List all business functions per user level as test steps

Figure 3: Business Process Test Script Example (Main Menu)

Acceptance Testing Action	Date	Initials
1. System Administrator logs on		
2. Is the correct module name/date/version # displayed at the top of the window?		
3. Are the appropriate buttons available (system admin has access to all)?		
4. Are the appropriate menu options available (system admin has access to all)		

Report Test Scripts

It is important that this process is tested for each module. If the result of printing is not as expected be sure to fill out a defect report form for that test.

It is important to repeat each test script for the different user level.

The procedures for testing the Reports are as follows:

1. Develop Report Test Scripts as shown in figure 4. Develop script for each user level you are testing.
2. Be sure to mark at the top of each Report Test Script which level you are testing.
3. Complete the test environment information at the top of each test script.
4. Login the appropriate User level.
5. For each step in the test script initial and date the successful tests.
6. If the test results are not as you expect please complete a defect form for that test step. (Do not initial and date unsuccessful test cases).
7. Please return a copy of all completed Test Scripts and Defect Forms as per the rules and timeframes described under that Section.

List all report functions as test steps

Figure 4: Report Test Script Example

Acceptance Testing Action	Date	Initials
1. Can you access the Report module from the main menu?		
2. Are all appropriate reports listed based on the Security of the user being tested?		
3. Can you access the Detailed Information Report from the list of available reports?		
4. Is the correct report name/date/version # displayed at the top of the window?		
5. Are the appropriate buttons available (based on the security matrix)?		

Defect Tracking Form Example:

Modify the user levels listed in the form to match your application

Add any key information you wish to have reported with defects

FIGURE 5: Defect Tracking Form

User Acceptance Testing DEFECT #
Application Defect Tracking FormTest case Step #:
Tester Name: Module ID: Date:
User Level tested: ❑ APPMANCRS_APPLICATION_MANAGER ❑ CLERK1CRS_PROCESSING_CLERKData Entry Functionality ❑ CLERK2CRS_PROCESSING_CLERKData Entry Functionality ❑ CLERK3CRS_PROCESSING_CLERKData Entry Functionality
Problem Severity (Check One): ❑ 1. System Crash (Data Loss) ❑ 2. System Crash (No Data Loss) ❑ 3. Incorrect Functionality (No Work Around) ❑ 4. Incorrect Functionality (Workaround) ❑ 5. Cosmetic

Can be Reproduced (Check One):

❏ (E) Every Time

❏ (S) Sometimes

❏ (O) Occasionally

(1x) Happened Once

Defect Summary Description (One sentence description of problem):

Defect Description: (please print)

❏ Screen Print/Error Message Attached

Steps to Reproduce:

Figure 6: Security and Integrity Requirements Definition Checklist

Date: _____
Interviewee:
Interviewer:

Process Name	Activity Description	Question	Response	Notes/ Comments
Order Management	Creation/ Maintenance	How will end users be authenticated so that they can create a new order?		
		What will happen if an end user fails the authentication process?		
		Who will be able to create and change orders using the eBusiness solution?		
	Analysis and Approval	Will there be specific business rules that apply to new customer orders that are executed on the eBusiness solution? If so, what are they		
		What will happen to the transaction if one of the business rules is not met?		
	Execution	If all the business rules are met, what happens then?		

Process Name	Activity Description	Question	Response	Notes/ Comments
		Will there be formal audit trails that are used to validate that a transaction has been successfully completed?		
	Integration	Once a transaction has been executed, what other areas of the organization will need access to its contents?		
		What information does the other areas of the organization require and how is the information provided?		
	Monitoring and Archival	How will the transactions that were executed on the eBusiness solution tracked/monitored?		
		What information about the transaction will be tracked/monitored and what format is it stored?		

Glossary

Acceptance
The formal process of accepting delivery of a product or a deliverable.

Activity Duration
Activity duration specifies the length of time (hours, days, weeks, months) that it takes to complete an activity. This information is optional in the data entry of an activity. Work flow (predecessor relationships) can be defined before durations are assigned. Activities with zero duration are considered milestones (milestone value of 1 to 94) or hammocks (milestone value of 95 to 99).

Applet
A small Java program that can be embedded in an HTML page. Applets differ from full-fledged Java applications in that they are not allowed to access certain resources on local computers, such as files and serial devices (modems, printers, etc.), and are prohibited from communicating with most other computers across a network. The current rule is that an applet can only make an Internet connection to the computer from which the applet was sent.

Architecture
The design of the organizational structure of a system, its communication rules, and system wide design and implementation guidelines. Architecture is also sometimes referred to as "system architecture," "design," "high-level design," and "top-level design." The term "architecture" can also refer to the architecture document.

B2B
Business-to-business commerce conducted over the Internet. May be classified in two ways:

- Infrastructure, which includes auction-solution software, content-management software and web-based commerce enablers.
- Markets, which consist of web sites where buyers and sellers come together to communicate, exchange ideas, advertise, bid in auctions, conduct transactions and coordinate inventory and fulfillment.

B2B integrates business operations with both suppliers and customers by streamlining and automating:

- Product information and transfer.
- Price negotiations.
- Order entry.
- Shipping data.
- Billing and collection cycles.
- Customer contact information/technical assistance.
- Surplus inventory auction.
- Spare parts replenishment.

See also Digital Marketplaces.

B2C

Business-to-consumer commerce conducted over the Internet. Describe linking consumers to commercial entities in one-way networks. Deals directly with buyers and creates benefits mainly for sellers. B2C has attracted the larger portion of media interest, but has already been eclipsed by B2B in terms of online sales.

Characteristics of successful B2C sites include:

- Community - development around topical issues, interests and lifestyles.
- Context - the ability to aggregate and navigate information, aiding customer choice.
- Commerce - the ability to fulfill transactions and provide billing services.
- Content - often a way to attract users to a site and make them stay.
- Customization - ability to deliver personalized content and customized products and services.
- Communication - servicing and communicating status to the customer via a number of channels.

See also Digital Marketplaces.

B2E

B2E is business-to-employee, an approach in which the focus of business is the employee, rather than the consumer (as it is in business-to-consumer, or B2C) or other businesses (as it is in business-to-business, or B2B). The B2E approach grew out of the ongoing shortage of information technology (IT) workers. In a broad sense, B2E encompasses everything that businesses do to attract and retain well-qualified staff in a competitive market, such as aggressive recruiting tactics, benefits, education opportunities, flexible hours, bonuses, and employee Empowerment strategy.

A B2E portal has three distinguishing characteristics:

- A single point of entry: one URL for everyone within an organization.
- A mixture of organization-specific and employee-defined components.
- The potential to be highly customized and easily altered to suit the particular employee.

Browser

A program used for searching or "browsing" the Internet. Microsoft Internet Explorer (IE) and Netscape Communicator are examples of browsers for the World Wide Web.

Browser Caching

To speed surfing, browsers store recently used pages on a users disk. If a site is revisited, browsers display pages from the disk instead of requesting them from the server. As a result, servers do not accurately count the number of times a page is viewed.

C2B

Concerns the financial interaction, initiated by a consumer, between a consumer and business. See also B2B, B2C, C2C, And Digital Marketplaces.

C2C

The financial interaction between nonbusiness entities using the web. Traditionally, C2C e-commerce has been conducted through both trading forums and intermediaries such as auctions, classified advertisements and collectibles shows. See also B2B, B2C, C2B.

CGI (Common Gateway Interface)

A set of rules describing how a web server communicates with software on the same machine and how the other piece of software (the "CGI program") talks to the web server. Any piece of software is a CGI program if it handles input and output according to the CGI standard. Usually a CGI program is a small program that takes data from a web server and uses it. For example, a CGI program would put form contents into an e-mail message or turn data into a database query.

Cookie

The most common meaning refers to an information piece sent by a web server to the web browser. The browser software saves the information and sends it back to the server whenever the browser makes additional requests from the server. Depending on the type of cookie used, and the browser's settings, the browser may accept or not accept the cookie, and may save the cookie for a variable amount of time. Cookies might contain log in or registration information, online "shopping cart" information, user preferences, etc. When a server receives a request from a browser that includes a cookie, the server is able to use the information stored in the cookie. For example, the server might customize what is sent back to the user or keep a log of a particular user's requests. Cookies are usually set to expire after a predetermined amount of time and are usually saved in memory until the browser software is closed, at which time they may be saved to disk if their "expire time" has not been reached. Cookies gather more information about a user.

Customization

The options available to customers to select product or service components to meet a specific need. The parts or components are common and available to all customers irrespective of the specific individual customer characteristics. Examples of customization include the ability to select specific components of electronic products, such as computers, or the ability to order specific variations on component parts, such as the colors, engine size, type of seats, etc.

Data Transport Layer

This is Layer 4 of the OSI reference model. This layer is responsible for reliable network communication between end nodes. The transport layer provides mechanisms for the establishment, maintenance, and termination of virtual circuits, transport fault detection and recovery, and information flow control. Corresponds to the transmission control layer of the SNA model.

Data Warehousing Management

The on-going supervision of the data warehousing process. A data warehouse collects, organizes and makes data available for the purpose of analysis. It gives management the ability to access and analyze information about its business.

Digital Marketplaces

Enables multibuyer and multiseller interaction and collaboration. Comprised of industry specific search engines, information marts and business malls. Value is added by digital marketplaces through the aggregation of buyers and sellers, creating liquidity by generating a critical mass of buyers and sellers and reducing transaction costs. In addition, they facilitate information and knowledge-sharing trading communities. In contrast to consumer hubs, the value created by digital marketplaces increases exponentially with each additional participant in the network. Digital marketplaces can use a variety of market-making mechanisms between participants to mediate transactions (catalogues, auctions and exchanges). The term digital marketplace encompasses that of infomediary. The primary source of revenue for digital marketplaces are transaction commissions, for infomediaries it is typically subscriptions. See also Infomediary, Vertical Digital Marketplaces, Horizontal Digital Marketplaces.

Digital Certificates

The electronic equivalent of an ID card, used in conjunction with a public key encryption system. Also called digital IDs, digital passports, X.509 certificates, or public key certificates.
A digital certificate is an owner's public key, which a certificate authority has digitally signed. Certification authorities (CAs) such as VeriSign, Inc., issue a digital certificate to an individual after verifying that a public key belongs to that individual. The certification process varies depending on the certificate authority and the level of certification. The process may require identification such as drivers licenses, notarization, or fingerprints.

Deliverable

Any work product that must be delivered to someone other than the work product's author

Deployment

The procedure to put a new system in to operation, especially what must be done for a new system to replace older system.

Design

The process of conceiving, inventing, or contriving a scheme for turning a specification for a computer program in to an operational program; the activity that links requirements development to construction. "Design" also refers to the result of the design activity.

EBusiness
Transforms the exchange of goods, services, information and knowledge through the use of network enabled technology.

EC
(Electronic Commerce or e-commerce) The use of network-enabled technology to initiate and record sales and purchase transactions. See eBusiness.

EDI
(Electronic Data Interchange) Traditionally used in electronic commerce. A system used to electronically exchange information like purchase orders, acknowledgements, invoices and other routine business documents.

EFT
Electronic Funds Transfer.

ERP
(Enterprise Resource Planning) Software that integrates all departments and functions across a company onto a single computer system. Examples would be the PeopleSoft and SAP software packages.

Firewall
A filter between a corporate network and the Internet that keeps the corporate network secure from intruders, but allows authenticated corporate users uninhibited access.

Gateway
The technical meaning is a hardware or software set-up that translates between two dissimilar protocols. Another meaning is to describe any mechanism for providing access to another system.

GUI
Graphical User Interface.

Internet
A worldwide system of computer networks. Conceived by the Advanced Research Projects Agency (ARPA) of the U.S. government in 1969 and first known as the ARPANet. The original aim was to create a network that would allow users of a research computer at one university to be able to communicate with research computers at other universities.

Because messages could be routed or rerouted in more than one direction, the network could continue to function even if parts were destroyed in the event of a military attack or other disaster.

Intranet
The subset of the Internet used internally by a company or organization. Unlike the Internet, intranets are private and accessible only from within the organization. Prior to Internet technology, most corporations relied on proprietary hardware and software systems to network computers. Now, Internet technology allows for internal communication where information is stored on company servers and accessed through a web browser. In essence, an intranet is a miniature Internet using many of the same features, such as individual home pages, newsgroups, and e-mail. See also extranet.

Load Balancing
The process by which load (number of requests, number of users, etc.) is spread throughout a network so that no individual device becomes overwhelmed by too much traffic, causing it to fail. Load balancing also involves redirection in the case of server or device failure to allow for Failover and promote Fault tolerance.

Market Integration
An approach to business consulting, which focuses on helping companies to develop new and innovative relationships between their customers, suppliers, investors and employees. (e.g. Success in the new economy requires organizations to simultaneous address business issues across a complex set of internal and external markets.)

NAT
Network Address Translation (NAT)-A method used to connect multiple computers to the Internet using one IP address. NAT is often used as a security defense to protect true IP address of a web server that is behind a firewall.

Packet Sniffer
A tool used to intercept data as it travels across the Internet. It is often used by hackers to steal passwords and information.

PKI (Public Key Infrastructure)
Enables users of a basically open public network to securely and privately exchange data and money. PKI utilizes a cryptographic key pair

that is obtained and shared through a trusted authority. The public key infrastructure provides digital certificates and can identify individuals, organizations and directory services that can store and, when necessary, revoke them. Although the components of a PKI are generally understood, a number of different vendor approaches and services are emerging. The public key infrastructure uses public key cryptography, the most common method on the Internet for authenticating a message sender or encrypting and decrypting a message. Traditional cryptography has usually involved the creation and sharing of a secret key for the encryption and decryption of messages. If the private-key system is intercepted, messages can easily be decrypted. For this reason, public key cryptography and the public key infrastructure is the preferred encryption method on the Internet. The private-key system is sometimes known as symmetric cryptography and the public-key system as asymmetric cryptography.

Plug-in

A software piece that adds features to a larger piece of software. Common examples are plug-ins for a web browser. Plug-ins are loaded into memory by a larger program to add new features. Users need only to install the few plug-ins that they need out of a much larger pool of possibilities.

Portal

A web site that is the first place people visit when using the web. Typically a portal site has a catalog of web sites, a search engine or both. A portal site may also offer e-mail and other services to entice people to use that site as the main point of entry or portal to the web. Portals are designed to be the "front door" through which a user accesses links to relevant web sites. Information categories typically include shopping, browse, classified ads, read current news, get weather/stock updates, etc.

Public Key

Value provided by a designated authority that can be used to effectively encrypt and decrypt messages and digital signatures The use of combined public and private keys is known as asymmetric cryptography. A system for using public keys is called a public key infrastructure (PKI).

Router

A special-purpose computer or software package that handles the connection between two or more networks. Routers look at the destination addresses of the packets and determine routing.

Search Engine
Creates databases or indexes of Internet sites based on titles, key words or the full text of files. When searching, a list of results is presented. Web-based search engines present results in hypertext, allowing users to click on the results and retrieve the website. If the search results are unsatisfactory, the "back" button on the browser returns users to the search. Some browsers allow users to bookmark the results of a search and refer to the list at a later time. A user with a personal web site can register the site with all search engines and submit key information. This information is added to the index so other users can receive the location in answer to a search.

Security Certificate
Information, stored as a text file, that is used to establish a secure connection in a network. Security certificates contain information about the owner, the issuer, a unique serial number or other unique identification, valid dates and an encrypted "fingerprint" that can be used to verify the contents of the certificate. In order for an secure connection to be created both sides must have a valid security certificate.

Server
A computer, or software package, that provides a specific kind of service to client software. The term can refer to a particular piece of software, such as a World Wide Web server, or to the machine on which the software is running.

SET
(Secure Electronic Transactions) A system for ensuring the security of financial transactions on the Internet. It was initially supported by MasterCard, Visa, Microsoft, Netscape, and others. With SET, a user is given an electronic wallet (digital certificate) and a transaction is conducted and verified using a combination of digital certificates and digital signatures among the purchaser, merchant and the purchaser's bank. The transaction is conducted in such a way that it ensures privacy and confidentiality. SET makes use of Netscape's Secure Sockets Layer (SSL), Microsoft's Secure Transaction Technology (STT), and Terisa System's Secure Hypertext Transfer Protocol (S-HTTP). SET uses some but not all aspects of a public key infrastructure (PKI).

SMTP

(Simple Mail Transfer Protocol) The main protocol used to send electronic mail on the Internet. SMTP consists of a set of rules for how a program sending mail and a program receiving mail should interact. Almost all Internet e-mail is sent and received by clients and servers using SMTP, thus if one wanted to set up an e-mail server on the Internet one would look for email server software that supports SMTP.

SSL

(Secure Sockets Layer) A program layer created by Netscape for managing the security of message transmissions in a network. Netscape's idea is that the programming for keeping your messages confidential should be contained in a program layer between an application (such as your web browser or HTTP) and the Internet's TCP/IP layers. Sockets refer to the method of passing data back and forth between a client and a server program in a network or between program layers in the same computer. Netscape's SSL uses the public and private key encryption system from RSA, which also includes the use of a digital certificate. SSL is an integral part of each Netscape browser. If a web site is on a Netscape server, SSL can be enabled and specific web pages can be identified as requiring SSL access. Other servers can be enabled by using Netscape's SSLRef program library, which can be downloaded for noncommercial use or licensed for commercial use.

TCP/IP

(Transmission Control Protocol / Internet Protocol) A language that governs the communication between all computers on the Internet. TCP/IP is a programmed set of instructions that dictates how packets of information are sent across multiple networks. The Transmission Control Protocol ensures that the packets arrive correctly at their destination address. The Internet Protocol determines where packets are routed, based on their destination address. In order to have full Internet access, users must be running TCP/IP. Related term: PPP.

URL

(Universal Resource Locator) The address for a web site, directory or file on the World Wide Web. URL is also a convention that web browsers use for locating files and other remote services. Related term: IP Address.

VAN

Value Added Network

VAR

(Value Added Reseller) A company that takes an existing product and adds value usually in the form of a specific application for the product. The company then resells the product as a new product or package.

Vortal

A web site that provides a gateway or portal to information related to a particular industry. A term that might also be used is interest community web site, since any vertical industry joins people that share an interest in buying, selling or exchanging information about that industry. Vortals are also seen as likely business-to-business communities. For example, small business people with home offices might be attracted to a comprehensive vortal that provided ideas and product information related to setting up and maintaining the home office. A vortal gives the user a single place to communicate with and about a single industry.

VPN

(Virtual Private Network) Refers to a network in which the parts are connected using the Internet. The data sent across the Internet is encrypted, so the entire network is virtually private. A typical example would be a company network where there are two offices in different cities. Using the Internet the two offices merge their networks into one network, but encrypt traffic that uses the Internet link.

WAN

(Wide Area Network) Any Internet or network that covers an area larger than a single building or campus. See also LAN.

WAP

(Wireless Application Protocol) Specification for a set of communication protocols to standardize the way that wireless devices, such as cellular telephones and radio transceivers, can be used for Internet access. WAP includes e-mail, the World Wide Web, newsgroups and Internet Relay Chat (IRC).

WML

(Wireless Mark-up Language) A mark-up language providing a 'light' version of a web site for viewing on handheld devices.

WWW

(World Wide Web) The resources and users on the Internet that are using the Hypertext Transfer Protocol (HTTP). A broader definition comes from the organization that web-inventor Tim Berners-Lee helped found, the World Wide Web Consortium: "The World Wide Web is the universe of network-accessible information, an embodiment of human knowledge."

XML
(Extensible Markup Language) A flexible way to create common information formats and share both the format and the data on the World Wide Web, intranets and elsewhere. Any individual or group that needs to share consistent information can use XML.

X509
A specific format used for digital certificates. It is used in public key infrastructure architecture to distribute an end user's public key.

Index

About the Authors

Wes Balakian PMP, CEA, brings a wide range of project management and eBusiness experience in managing software development projects. With more than tens years of global program management experience in the telecommunications industry. Current responsibility focuses on implementation and management of complex, medium to large-scale, business to business solutions for Internet commerce, supply chain management, client relationship management, workforce optimization and industry-specific offerings.

Mr. Balakian is also the president of True Solutions Inc. TSI was created to further develop the need for education, training and development in the IT eBusiness and e-commerce project management profession. As an accomplished speaker, Wes has delivered eBusiness project management training to fortune 100 companies in six continents. Other responsibilities include: the Vice Chairman of Professional development for the PMI EBusiness SIG (Specific Interest Group), Vice Chairman at large for the B2B Technology Group.

Keith Young CISA, CePM, is the Managing Partner of Integrity Solution, LLC. He has over 9 years experience in assisting companies in the health care, manufacturing, governmental, telecommunications and financial service markets in implementing ERP and eBusiness solutions. During his career, Mr. Young has trained and consulted with several Fortune 500 companies on eBusiness concepts and strategies. Some of his clients include the Federal Reserve Bank, BellSouth, General Motors Corporation and Shell Oil.

He is the Vice Chair of Marketing for PMI's EBusiness SIG (www.pmi-eBusiness-sig.org), a global organization made up of over 1,000 members that are involved in the development of eBusiness project management best practices. He is also the founder and chairperson of the B2B Technology User Group. (www.b2btechnologyusergroup.org), a global think-tank organization made up of over 300 members that are involved in the development of best practices for designing, implementing and supporting business to business (B2B) enterprise solutions.

Rajesh Veerapaneni has worked as a Systems Analyst, Systems Manager, and Systems Architect for large corporations. With strong technical background and hands on experience in large ERP and eBusiness implementations. Currently, working as Manager SAP BASIS and Infrastructure for a global supply chain management firm whose clients are HP, Microsoft, IBM and Compaq.

Rajesh is a Computer Engineer and spent more than a decade in the software profession with good project management experience in large-scale ERP/ eBusiness projects. As an accomplished speaker, Rajesh has delivered ERP, eBusiness project management training in USA, India and Germany. Other responsibilities include: the Vice Chair of Technology for the B2B Technology Group.

www.ingramcontent.com/pod-product-compliance
Lightning Source LLC
Chambersburg PA
CBHW030937180526
45163CB00002B/601